I0762570

MYTHS, LEGENDS AND TALES FOR A GREENER WORLD

This book is dedicated to everyone and all living creatures in the world. With grateful thanks to the numerous unknown people who first told these stories, through time and across the globe, and to the many others who kept them alive through the ages by retelling, transcribing and translating them.

First published in the United Kingdom
in 2026 by Batsford
43 Great Ormond Street
London, WC1N 3HZ

An imprint of B. T. Batsford Holdings Limited

Illustrations by Joe McLaren

ISBN 978 1 84994 998 9

A CIP catalogue record for this book is available from the British Library.

10 9 8 7 6 5 4 3 2 1

Reproduction by Rival Colour Ltd, UK
Printed by Vivar Printing Sdn. Bhd., Malaysia

This book can be ordered direct from the publisher at www.batsfordbooks.com, or try your local bookshop

Distributed throughout the UK and Europe by Abrams & Chronicle Books, 1st Floor, 22–24 Ely Place, London EC1N 6TE and 57 rue Gaston Tessier, 75166 Paris, France

www.abramsandchronicle.co.uk
info@abramsandchronicle.co.uk

MIX
Paper | Supporting responsible forestry
FSC® C084469

ROSALIND KERVEN

MYTHS, LEGENDS AND TALES FOR A GREENER WORLD

A BOOK OF GLOBAL WISDOM

Illustrated by Joe McLaren

BATSFORD

CONTENTS

The World and its Stories

In every country in the world, and within every culture, a set of myths and legends form part of time-honoured tradition.

No one knows who originally composed these stories; most began by being told aloud. They were then passed on to other people in the same way, developed and enhanced over many years by countless unknown retellers. Some were written down in antiquity, but others continued to be shared only by word of mouth until very recent times.

The stories and fragments of wisdom presented here come from across the globe: Africa, the Americas, Asia, Europe, the Middle East and the Pacific islands. They are drawn from bygone civilizations, modern indigenous peoples and recent Western culture. They have crossed borders, continents and seas – north, south, east, west and back to the very centre. Despite their very different roots, all contain universal truths and insights about the problems currently threatening our beautiful planet Earth and the myriad creatures that share it with us.

For, regardless of where and how we live, we are all in this together.

WHERE DID WE COME FROM,
HOW DID WE GET HERE?

Beginning

Tibet

At first there was only the void:
pure emptiness.
Nothing existed.

In the infinite stillness, light glowed.
Time began.

It passed, unfathomably slowly.

The light split into two,
forming Gloom and Radiance –
waiting to act …

Turmoil! Chaos!

Endless streams of colour – flashing, coiling, lengthening, shrinking, tearing, mingling … Then gradually separating into five things:

Hardness
Fluidity
Heat
Motion
and Space.

The five merged into one,
becoming a huge egg
that split into two halves.

From one half, Gloom drew out evil demons.
Their names were:
Disease
Misfortune
Drought
Hunger
and Despair.

But from the other half, shining brightly, Radiance drew out benevolent gods.

They were called:
Virtue
Good Fortune
Prosperity
Health
and Joy.

Creation

Cheyenne people (Native American)

Through the Universe moved All Spirit, bursting with creative power.

All Spirit thought of water. At once, a great expanse of water came into existence, salty and cool.

All Spirit thought of living beings.

Fish appeared in the water, alongside other creatures: shellfish crawling on the sandy bottom, ducks and geese skimming the surface.

The air filled with sound: water lapping, wings flapping, feet splashing. But nothing was visible, for it was dark.

So All Spirit wished for light to see by.

Light grew, thin and pale at first, then spreading, turning golden, revealing the new world in all its beauty.

Time went by.

The newly created beings found that they too had thoughts and desires. So Snow Goose approached All Spirit on behalf of them all.

'Powerful one,' she honked, 'We feel your presence and thank you for the four marvellous things you have given us: our lives, water, light and the sky. But we birds long to see what lies beyond.'

'Since you make this request with such courtesy,' All Spirit replied, 'I grant you the skill of flying.'

At these words, all the birds suddenly felt their wings quivering with energy. In great excitement, they rose from the surface of the water, up into the air. They flew here and they flew there, tracing exquisite patterns through the sky as they explored it, moving far, far away in all directions.

When they eventually returned, Loon spoke up in his high pitched, plaintive voice: 'All Spirit! Again we give thanks for everything you have made and all the wonders you have performed. Please forgive us for requesting more, but we birds are exhausted from endlessly swimming and flying. Would it be possible to give us a dry, solid place where we could build nests and sleep?'

'It is possible indeed,' said All Spirit. 'But creating so many things already has drained away much of my power. I cannot make the kind of place you desire without assistance.'

Loon gazed round at his fellow creatures. They answered in a chorus of voices:

'We will all help!'

'Tell us how!'

'We will do our best!'

All Spirit answered: 'Thank you. There is something hidden, far below the water. It is called "land". I need one of you to go in search of it.'

Snow Goose honked, 'I volunteer to try!'

She set out at once, gliding eagerly across the water, faster and faster, then soaring up into the sky with a thunderous beating of wings. Higher and higher she went until she could only be seen as a dark speck against the light. Then she plunged down into the water like a spear, dropping swiftly through the depths.

While the other birds waited, they passed the time by counting. They had just reached four times 400 when Snow Goose finally emerged at the surface, gasping for air.

'I found nothing,' she said. 'I am so sorry.'

All Spirit praised her great effort and thanked her.

The next attempt was made by Loon. After him went Mallard Duck. Each flew to the furthest heights, then plummeted into the depths just as Snow Goose had done. They too had no success.

Then a fourth bird, little Coot, came paddling through the water, every so often dipping his head to catch a fish. 'All Spirit,' he said softly, 'I have something to tell you. Whenever I put my head below the surface, I see something dark and mysterious far, far below. Could it possibly be this "land" that we are seeking? I cannot fly high or dive quickly like my larger sisters and brothers here; but I could try swimming slowly down to investigate this mysterious thing, if that would help?'

'Little brother,' said All Spirit. 'Each of you has different talents, and no one is more or less capable than the others. Perhaps your slow approach may prove more successful than hasty diving. If you have the courage, I urge you to try.'

Coot grunted, put his head under the water and swam gradually down, down and down. A very long time passed. Then … there he was, rising back up. When he reached the surface, he did not speak or gasp for air like the others. For his beak was tightly shut – holding something.

'Well done, little brother,' said All Spirit – and suddenly materialized into human shape, the shape of a man.

In this form, he squatted on the surface of the water, holding out one hand. Coot opened his beak and dropped a tiny ball of mud onto the open palm.

All Spirit used both his hands to roll and knead the mud. As he did so, it gradually grew larger. And larger. Eventually, it was too enormous for him to hold.

So All Spirit said, 'This mud must be placed somewhere permanent. But it is too heavy to float in the air or drift on the water. Can any of you water beings carry it on your back?'

They all came crowding around eagerly.

First, the shellfish cried in unison, 'Look how solid our backs are! We can carry the mud between us.'

'Thank you, sisters and brothers,' said All Spirit, 'but you live much too far down in the depths.'

In their wake came fish of many sizes and colours, wriggling eagerly. 'Try us!' they called.

All Spirit said, 'I am immensely grateful, my friends. But sadly, your backs are too narrow, and your dorsal fins would cut through the mud and make it disintegrate.'

That seemed to leave only the birds. But they were all exhausted from constantly swimming and flying. Was there no creature suitable for this monumental task?

Fortunately, there was: someone with a wide, solid back – someone who liked to be still, resting just below the surface. She swam forward, graceful as a drifting shadow, raised her wrinkled head and called out softly, 'All Spirit, though we are still in the dawn of the world, I have already grown strong and steady. Let me take this burden.'

'Grandmother Turtle,' said All Spirit. 'Thank you. I lay this mud on you as a sacred gift.'

'It is a great honour to receive it,' said she.

All Spirit placed the mud onto Grandmother Turtle's back, spreading it across her, completely covering her all over, until she could no longer be seen.

All Spirit said, 'You shall have a reward for accepting this mighty burden, Grandmother Turtle. I promise that you and your descendants shall always be equally at home, not just underwater, but also within the earth and above the ground. In this way, the whole world will be yours.'

After these words, All Spirit's human form melted away.
Now there was only his invisible presence, and silence.
All the creatures swam and flew and rested on the mud.

More time passed.

All Spirit thought of plants.

At once, the mud was rooted with grass, trees and flowers, gracing it with the beauty of natural growth, transforming it into Earth Woman. The flowers matured into fruits and seeds, scattering across her body, making even more plants grow.

The various creatures were now very content. The birds rested on Earth Woman when they were tired. The fish swam closely around her sides. Her love could be felt everywhere.

But sometimes, when the wind passed over her, she sighed – for she was lonely.

So All Spirit thought of bones.

At once, two bones came into being. All Spirit laid them softly on Earth Woman's body and breathed on them. The bones came to life: a woman and a man – the first two people.

All Spirit created many other land animals to live alongside them, to help fulfil their needs for food, clothing and shelter. The first people lived happily together. They had many children, who matured and spread around the world. These were the ancestors of all the different groups and nations that we know today.

And today, All Spirit still watches over Earth Woman and her countless children – not just as their creator, but also as the guardian of life.

Growth

Tahiti

There were tens of roots.
There were hundreds of roots.
There were thousands of roots.
There were myriads of roots …

Rain increased. It fell everywhere.
Moss and slime grew and increased.
Forests grew and increased.
Food grew and increased.
Creeping plants grew and increased.
Living things grew
In the sea and the rivers and the land.

Seasons

Scotland

It was still early times when Old Bone Mother came into being. Grey-faced and frosty-haired she was, harsh as a storm at sea. She spread her power across much of the world, blighting it with permanent winter.

Cold she made it, barren and bleak. How all the people and animals suffered!

Old Bone Mother had a son in the prime of youth. He had no share in her dominion.

One night this youth dreamed of a sunlit lady. He awoke smitten with desire to find her. He went to ask his mother for help, since she knew the answer to everything. However, she only scowled at him witheringly, and sent him packing.

So, the youth went travelling in search of his dream lady, hither and thither, all over the land. He found no clues of her anywhere. He built a boat and set sail across the ocean, until he reached a distant island.

This island was ruled by a lord who willingly offered assistance. He told the youth, 'The sunlit lady you dreamed of is destined to be the queen of summer, with you beside her as the king.'

'Summer?' the youth asked. 'What is that?'

The island lord replied, 'Summer is the opposite to the evil that Old Bone Mother has inflicted on the world. It is warmth, sunshine and greenness. It is joy, fertility and pleasure.'

'That sounds like riches well worth questing for,' said the youth. 'But where can I find my dream lady? And what can we do to bring about this wonderful thing called summer?'

'The sunlit lady lives inside a mountain,' the island lord replied. 'This mountain itself is easy enough to find, but getting close to her will be a challenge. For your mother keeps her captive there, surrounded by impenetrable rocky walls, guarded by an impossibly heavy boulder.'

'Why has she created such a cruel prison?' asked the youth.

'Because Old Bone Mother fears allowing the sunlit lady to walk free by your side. If she did, together you could easily overthrow your mother's glacial power.'

The youth immediately vowed to carry out this deed.

He dressed himself in shining gold, sailed back across the ocean and rode out in search of the impenetrable mountain. Now he knew exactly what he was seeking, it did not take him long to locate it. As the next chill winter dawn streaked across the sky, he found the entrance. Summoning all his strength, bolstered by his yearning, he hauled away the huge boulder that blocked it, and helped the sunlit lady to step out.

In this way, the pair were united. They walked joyfully together across the wintry land. They spread their passion around to melt the frost, soften the soil and release the promise of germination. Each time they came to a frozen river, they gently thawed the ice, freeing a melodious surge of water. Birds burst into song, heralding their good work.

But Old Bone Mother saw what they were up to and flew into a rage.

She spat at the grass for its audacity in starting to grow. She mangled the buds before they could properly open. She battered her magic hammer against the ground, turning it ice-hard again.

Then she called up the north wind and the east wind, shrieking at them, 'Rise up and attack! Shake that brazen son of mine in tempests and hailstones! Beat him, bludgeon him, drive him away!'

The winds did her bidding. But they could not fully overcome the youth. For the following daybreak he returned, this time bringing winds of his own, from the south and the west. However, they were too mild to withstand Old Bone Mother's warriors, and again she drove him away. He came back at the next dawn – only to be beaten off once more.

Thus it went on, day after day, with the youth harnessing light and warm winds against Old Bone Mother's devastating storms.

Meanwhile, the sunlit lady fled over the hills, cowering in valleys, buffeted across oceans in the ruinous turmoil. How the world suffered and shivered alongside her!

But Old Bone Mother could not hold them back for ever. Ravaged by her own violence, she slowly began to weaken, and her power crumbled like dust.

New growth pushed against her unstoppably, smothering her fury in living green.

Feeble she grew, so feeble. Finally, she fled – spitting snow behind her, flicking sleet from her fingertips, tramping damp meadows into mud, tearing hesitant new leaves with gales …

Until her anger completely died away.

Now the living world and its creatures could renew themselves.
Soon they were flourishing once again.
Summer ruled!

It has been claimed that, in the end, Old Bone Mother turned to stone – but that is wrong. For after her defeat, she managed to drag herself secretly to the Well of Rejuvenation, and drank deeply of its magic waters. More months turned. On and on she drank. Gradually, the waters renewed her strength …

Until once again, she ravaged the world with bitter winter.

Thus the battle between the seasons continues until eternity. Each and every year, Old Bone Mother bewitches all the plants to slumber, frightens some creatures to hide in the warm bowels of the Earth, and burdens those that run free with the weight of thick, enveloping coats. She also keeps us people in check, alongside all the dormant plants and seeds.

For Old Bone Mother is an intrinsic part of the turning world, one of the great transformers.

THE WORLD IN TROUBLE

Good and Evil

Mapuche people (Chile and Argentina)

All around us there is constant conflict
between the forces of good and evil.

Good brings order and harmony,
benefitting human beings
and the entire natural world.

However, evil means uncertainty,
chaos and destruction,
so that all humans
and other living things
must suffer.

Man-made Climate Disaster

China

A young nobleman was fond of deer hunting. One summer he joined a group of fellow enthusiasts at a small mountain village, staying in simple but comfortable local lodgings. Their hosts served them very generous helpings of food. This was surprising, for a prolonged drought was badly affecting all the local farms, making the crops wither where they grew. The villagers themselves went hungry so they could feed their guests; for in those days, the humble always deferred to the powerful.

Late one afternoon as the hunters were returning to their lodgings, a huge stag suddenly appeared, sporting a splendid set of antlers. The young nobleman was entranced by it. As the stag sprung away up the hillside, he spurred his horse to follow it, leaving his companions behind. Up and down countless slopes they chased, to the edge of a thick, dark forest. Here the stag slipped into the trees with the young nobleman still racing after it on horseback. The stag leaped and pranced on the narrow track, always just too far ahead … until eventually, it vanished.

The young nobleman came to an abrupt halt, exhausted and sweating. The sun had long ago set, and it was almost dark; he could no longer see the track he had come along. He was completely lost. The uncanny silence was broken only by screeches of owls and monkeys. The forest stretched away in all directions without end, hostile and impenetrable.

Then, to his great relief, he noticed a light glimmering far away through the crowded trees. He urged his weary horse towards it and emerged into a clearing beside a wide lake. On its shores stood a grand mansion with towering white walls. The light he had spotted was shining from exquisitely ornamented bronze lanterns hanging over a red lacquered door. He ran to it and knocked loudly, calling for help.

The door was quickly opened by a male servant. The young nobleman explained his situation and asked if he could be given a bed for the night.

The servant looked him up and down. 'I will do what I can to help you, sir,' he said. 'However, both my masters are away at the moment, so I must seek permission from their mother, the dowager.'

He shortly returned with a groom to attend to the horse, and ushered the young nobleman into a grand hallway, its walls lined with red curtains and dark wood. There he was greeted by an elderly lady. She was straight backed and very elegant, wearing a richly embroidered silk gown. Her finely sculpted face was white as porcelain but dusted with wrinkles, and the hair piled on top of her head was silvery grey.

'Good evening, stranger,' she said. 'I hear you are lost in the forest. Well, I am willing to accommodate you until the morning, on two conditions. Firstly, since my two sons are absent, you must stay in your room and not leave it without permission. Secondly, if anything untoward happens during your stay, you must accept what you are told without question.'

Since the young nobleman had no choice, he agreed to these terms. The dowager led him into a dining room and bade him sit down. They were served a delicious meal, entirely comprised of fish on crystal platters. As soon as it was over, he was shown to a luxurious bed chamber. When the servant went out, the young nobleman heard a key turning, then removed from the lock. He was trapped!

He lay down and dozed nervously.

Around midnight, he was suddenly jolted wide awake by a loud banging at the main door, and a deep voice shouting urgently:

'Open up! By order of Heaven! We have an urgent order for rain to be delivered tonight, over a radius of 700 miles. No omissions are permitted! No unauthorized excess!'

He heard footsteps running and hinges creaking as the main door swung open. Then he recognized the voice of the servant who had admitted him earlier, saying, 'Please convey our humble apologies to the Emperor of Heaven! I regret that both my masters are thousands of miles away, so they cannot possibly fulfil this order tonight. Only the dowager is present, and she is far too frail to carry it out.'

'No excuses are acceptable,' the messenger shouted back. 'Tell your mistress she must find a way to obey the Emperor of Heaven. Failure to do so will result in severe punishment.'

The door slammed shut and the young nobleman heard no more. He leaped from the bed and straightened his clothes.

Soon he heard the muffled tones of the servant talking anxiously to the dowager.

Then her crisp voice replied, 'Let me think … Ah, yes! By lucky chance, there *is* a solution to this problem. That stranger staying with us seems both strong and capable. Bring him to me at once.'

Moments later, the young nobleman's door was unlocked and the servant was beckoning him out, crying, 'Hurry! The mistress wishes to speak with you.' He led the young nobleman to the entrance hall, where the dowager awaited him.

In the muted glow of night lights, she looked different from before.

Her diminutive body seemed to be flickering. A draught from the open window had blown her pile of silvery hair loose, making it drift around her face like a mane. The skirt of her dress had taken on a curious drape, resembling a tail. Her eyes were glowing eerily. The young nobleman blinked. Was he hallucinating?

She answered the question for him: 'Young man, I'm afraid I did not introduce myself properly when you arrived. However, now it is vital to reveal the truth: I am no ordinary lady, but the dragon king-mother of this realm. My two sons, who are currently absent, jointly rule from this palace as dragon kings.'

'So that is the meaning of the summons I just heard!' the young nobleman exclaimed. 'As dragon kings, your sons must be responsible for all the water in their realm, including the rain?'

'Indeed.' The dowager nodded solemnly. The constant flickering of her body between human and dragon forms was so disconcerting that the young nobleman had to avert his eyes. 'My elder son is at a wedding,' she went on, 'and his younger brother is escorting his sister on a dangerous journey. I told them it was ill-advised for both to be absent at the same time, but they refused to heed me. Now the Emperor of Heaven has ordered them to deliver some urgently needed rain across their lands. The only way to fulfil this order without them – is for *you* to do it!'

The young nobleman thought back to the parched fields around the village where he was staying, and his malnourished yet generous hosts.

'It sounds like an emergency,' he said. 'I will certainly try, madam. I will do my best.'

'I hope you have no fear of heights?' said the dragon king-mother. 'For you will need to ride my sons' rainmaking horse right up in the clouds.'

'I'm a competent horseman,' he replied, 'and not afraid to face a challenge. Please explain exactly what is required.'

'Wait here while I arrange everything, then I will tell you,' was her answer. She bustled off, calling servants as she went.

Soon she came back and ushered him outside. The night was very dark, lit by neither moon nor stars. Under the dim lantern glow, a groom appeared, leading a beautiful piebald stallion. There was something unearthly in both its eyes and its gait. At a word from the dowager, the young nobleman mounted it. The saddle was fashioned from fine leather, but the horse had no reins, not even a bridle. The dragon king-mother passed him a small jar with a tight-fitting lid, pierced with a small hole.

'This is the rain jar,' she said. 'Fasten it to that hook you see on the right side of the saddle. Good. Now then, let the horse choose his own path through the sky; do not attempt to guide him. Every so often, he will suddenly stop and whinny. That is your cue. Unhook the rain jar, turn it upside down and shake it once. That will release a drop of water from it, directly onto his mane. Shortly afterwards, you will see a good rain shower falling onto the land below.'

'Surely a single drop isn't enough to make rain?' the young nobleman asked, incredulously.

'It is exactly the right amount,' she corrected him. 'Do you not remember, young man? When you arrived, I told you not to question anything. Now then, listen carefully and remember well. Release just one drop, and one drop only, each time. *Do not ever shake out any more* – or you will cause a terrible disaster.'

The young nobleman nodded and saluted her.

She hissed a mysterious word under her breath … and at once, the piebald stallion set off, quickly breaking into a gallop, then rising into the air like a bird! Though the young nobleman had no reins to steady him, the horse moved so smoothly that he easily kept his place in the saddle. Soon he was greatly enjoying the ride.

Higher and higher they rose through the darkness, passing through dense billows of clouds, then emerging above them into the starry firmament. On and on they galloped, flying through the wind.

Without any warning, the horse suddenly stopped short and let out a long 'Neighhh … hh!'

The time had come! The young nobleman carefully pulled the jar free. Taking care to hold it securely, he inverted it over the horse's mane, shook it once, then quickly returned the jar to its hook. The drop that fell from it gleamed like a translucent pearl. The next moment, the clouds below him dissolved and he heard the sweet sound of gentle, steady rain falling onto the land far below.

The horse set off once more across the sky. Eventually, it again stopped and whinnied. Again, the young nobleman shook out a single drop from the jar, and again rain at once began to fall. In this way they continued, until the young nobleman considered himself as skilled at rainmaking as any dragon king.

Unfortunately, he was the kind of man who lets success go to his head. His satisfaction at performing the task well quickly turned to pride. Pride gave way to arrogance. This in turn transformed to sheer recklessness.

At the horse's next stop, the young nobleman saw through a gap in the clouds that they were poised immediately above the very village where he and his friends were staying. There were its arid brown fields full of wilted crops, and the dried-up river bed.

'The dragon king-mother's rule of one drop at a time is ridiculously restrictive,' he thought. 'She obviously has no idea how much rain is desperately needed down there. I shall use my initiative, and make sure they get a really good downpour, enough to properly revive their farms.'

He unhooked the jar, held it over the horse's mane and shook out – not one drop – but *20*.

At once, there came a deafening clap of thunder and a blinding flash of lightning – followed by the drumming of torrential rain.

'Excellent!' the young nobleman exclaimed out loud.

The stallion seemed to sense what he had done. It hesitated, trembled violently … then reared up so high that the young nobleman had to wrap his arms around its neck to keep his balance. It spun round and went galloping back through the sky, stopping no more until they had descended and landed in front of the dragon palace.

The dragon king-mother was already outside, waiting for them. The young nobleman leaped to the ground and ran to greet her, anticipating great praise for more than fully achieving his task.

However, she backed away and waved her arms at him angrily.

'Oh, you swaggering fool!' she cried. 'Why did you disobey the rules? Do you dare to think yourself cleverer than the divine dragon kings themselves? Ever since the dawn of time, they have carefully ensured that only the correct amount of rain falls on the Earth.

'I told you, I repeated it, and you confirmed that you understood: "Never shake out more than one drop at a time." That is an unbreakable, eternal rule established by Heaven itself. Yet at your last stop, you – a *lowly, ignorant mortal!* – took it upon yourself to shake out 20 drops! Yes, 20! Are mortals too foolish to understand? A single drop of rain brings water one foot deep. So 20 drops bring 20 feet of water – deep enough to submerge two men, plus two women all standing on each other's shoulders! You have completely flooded the village where this downpour fell. None of the people there can possibly escape alive!'

Global Warming

Cherokee people (Native American)

Every day, the Sun makes a long journey right across the arched roof that stretches above us – what we call the sky. As she travels, she gazes down on our world below, keeping an eye on what's happening here.

One day, she noticed something that made her fly into a rage. She saw that whenever people looked in her direction, they screwed up their eyes, making themselves appear grotesque. Everyone did it: men and women, rich and poor, weak and powerful. The Sun could not think of any explanation for this. So she concluded it must be to laugh at her. What an insult! What a lack of respect!

She couldn't stop thinking about it. She got angrier and angrier.

So she stormed across to the house where her brother, the Moon, lived and shook him awake from his daytime sleep.

'Hey, brother!' she cried. 'Have you seen the obnoxious behaviour of those people down on the Earth? Whenever I shine on them, they refuse to meet my eye and twist their faces into hideous grimaces. They're so bad mannered, so ungrateful! Why don't they appreciate all the light and warmth I give them? They should be getting down on their knees to thank me for my generosity.'

The Moon tried to placate her. 'They're not grimacing deliberately, sister,' he said. 'They can't help it. You see, Sun, your light is so dazzling, they're afraid you'll blind them. You should follow my good example: subdue your rays, make them softer and gentler.'

'I'd be ashamed to shine as feebly as you do,' the Sun retorted.

'Ashamed?' said the Moon. 'There's nothing shameful in people smiling up at me, admiring my beautiful, mellow light.'

This remark just inflamed the Sun's anger, for now she was jealous as well.

'Pah!' she cried. 'How dare they act as if your stingy, pathetic moonbeams were superior to *my* splendid rays! They've obviously forgotten that I am the most magnificent being in all creation. Well, since they are too conceited to admire me, I shall have to punish them.'

'Be careful, sister,' the Moon advised her.

'Careful?' scoffed the Sun. 'Not me! I've had enough of all the scowling and complaints. I'm going to burn all those people to death, one by one, until none of them are left.'

The Sun was as good as her word. From that time on, before she set out to travel across the sky each day, she fired up her rays to make them even hotter and brighter than ever before. They were scorching hot, blindingly bright – worse than anyone had previously imagined possible.

Then, as she made her usual journey across the sky, she pushed these scorching, blinding rays down to Earth with all her might. She thrust them through open doors into the houses, chasing away every last shadow, turning them unbearably stuffy. She made them penetrate even the thickest forests, so that all the wild creatures who normally found shade under the trees were left panting and gasping for breath. She forced her rays onto the rivers and lakes, drying up all the water there, leaving bare hollows. In this way, she enveloped the entire world in sweltering, scorching weather.

There was no escape from it. If clouds tried to resist her and provide some relief below, she callously pushed them aside. She banned rain; she banished cooling winds.

The leaves shrivelled up, the flowers wilted and the soil turned to dust. All the wild creatures suffered terribly.

The people grumbled and kicked up a fuss. They yelled out complaints to the Sun, their protests growing louder every day.

How the Sun laughed when she heard them! Her cackles tore through the air in rumbles of thunder. 'Look down there, Moon,' she called. 'I've given those insolent human beings exactly what they deserve!'

The Moon shrank away from her in disgust, not deigning to argue any further.

People began to fall ill. At first it was just a few of them, but soon there were thousands of victims. They became dizzy and disorientated from the torrid heat, parched with thirst, sweating with terrible fevers. No medicinal herbs could heal them, because the only possible cure was an end to the oppressive weather – and there was no sign of that ever happening.

Surrounded on all sides by the decaying corpses of wild creatures, one by one, people started dropping dead.

The remaining survivors were terrified that soon none of them would be left alive. So they got together in a council. Speaking in frail and shaky voices, they discussed their options. How could they stop the entire human race – alongside all the animals, birds and vegetation – from dying out?

The answer started as diffident mutterings. The mutterings grew louder into a clamour.

At last, someone declared: 'Friends, there is only one possible solution. We will have to kill the Sun!'

Wildfire

Hawaii

He comes quite gently at first ... then suddenly bursts into blinding flame. He races across the land, burning and consuming everything and everyone he meets.

They call him the Devourer of Trees, for the forest is his favourite target. As soon as his sparks touch them, the trees catch light. Smoke fills the air. Leaves tremble and shrivel as flames roll through them, unstoppable as a tsunami. Even the mightiest trees collapse: noble giants crumble to ash and cinders, overcome by suffocating heat.

His greed is insatiable; no matter how much he eats, nothing will ever quench his desire for more.

Everyone both respects and fears him.

He brews his fire in deep pits, nurtured secretly, deep under the ground. As soon as it overflows, reeking like rotten eggs, he hurls it out.

He doesn't stop at the forests, but also overwhelms whole villages. His flames engulf houses, gardens and fields, treasured places where people work tirelessly, attempting to produce bountiful crops. The Devourer mocks their labours, swallows them up and transforms their plantations to barren wastelands.

And what does he leave behind him? Nothing good.
Only:
choking clouds,
the sickening stench of death,
morsels of charred flesh, both animal and human,
terror and despair.

Nothing and no one is safe from him.

Drought

Somalia

In days of old, when poets daily declaimed their verses in marble palaces and citadels, there lived a king of the savannah who wished always to be prepared for danger. So, at the start of each new year, he would summon a soothsayer and ask, 'Will there be good luck or bad luck in the months ahead?'

One year the soothsayer might say: 'Majesty, your realm will be attacked by enemies, so prepare for war.' Another year, it might be: 'Plague is coming, so increase your stocks of medicinal plants.'

His prophesies were unfailingly correct. Thanks to his advice, the king kept his realm totally secure, thus enhancing his popularity.

However, one year the soothsayer warned, 'Mighty one, I foresee a worse problem than you have ever confronted before. There will soon be a terrible drought.'

'And how shall I overcome it?' asked the king.

The soothsayer shook his head. 'It will not be easy. This drought will be devastating. There will be not just months, but *years* of dryness, and countless deaths. Most of your population will perish long before the rains finally return. If you wish to survive and also maintain your power, you must be both clear headed and cunning.'

With those grim words, he snatched up his payment and hurried away.

The king pondered this alarming prophesy, then quickly summoned his slaves. 'Gather all the spare water and food you can find in every corner of the realm,' he ordered. 'I shall show you a remote and secret mountain hideout to store it in. Tell no one where this is, on pain of death.'

And there we must leave him for a while. But before we do, it must be revealed that this king was not a man – but a lion.

Just as the soothsayer had predicted, it permanently stopped raining.

Until then, the realm had been green and beautiful, graced with wide rivers, crystal clear streams and gleaming pools. But now all these waterways dried up into bare ditches and hollows, filled with only cracked mud and dust.

How could any living being survive such a situation? Only by creeping feebly around, desperately seeking the remnants of any springs that still trickled from the rocks, slow drip by drip. At such places, wild predators and their prey stood side by side, torpidly licking up whatever they could reach, too weak to attack or struggle. The people suffered too. Emaciated women sought water alongside the animals, scarcely managing to cover the bottom of their pitchers, cursing if they spilt even a single drop.

Another year passed. Still no rain came.

On the wide plains, from one horizon to the other, the grass withered into dead, brown mats. Even the deepest-rooted trees found no sustenance, permanently shedding their last leaves, buds shrivelling up. They dotted the parched land like wretched, petrified giants.

Sand blew up into blinding storms. The once lush and bustling savannah was now a desert.

The people no longer even tried to grow crops. Their herds of cattle, goats and sheep stopped yielding milk, then dropped dead of hunger and thirst alongside their owners. The camels lasted a while longer, until their humps shrivelled, and they too wasted away.

A third year passed.

The gazelles, antelopes, giraffes and other long-legged ones all fled over the mountains. The crocodiles snapped their deadly jaws no more, flies feasting on their desiccated corpses.

However, a handful of various animals stayed in their ancestral homeland, somehow clinging to life. They united to form a single band, lying low and taking turns to sniff out precious droplets of water in unlikely hidden places. These were the small ones – Gerbil, Mole and Shrew – and the clever ones – Serval the wild cat, Hyena, Baboon and Jackal.

One day, all these animals were slumbering in the sparse shadows of a dead thorn tree, when they were suddenly jolted awake by a bellowing roar.

'It's coming from those distant mountains,' said Hyena.

'It sounds like King Lion,' grunted Baboon. 'What a surprise – I thought he was dead.'

'He sounds in much better health than any of us,' said Mole. 'His roar is louder and stronger than ever.'

'He's probably sitting on a secret hoard of water and food,' mewed Serval. 'Wherever he's hiding, he'll have it well guarded to ensure no one else can get at it.'

'Then why isn't he keeping it secret?' squeaked Shrew.

'Perhaps he doesn't want to any more,' said Gerbil hopefully. 'He's a very noble animal, as we all know, so he may have realized it's his duty to help the rest of us.'

'Duty?' said Jackal. 'More likely, he's worried that if we all die, he'll have no subjects left to rule over – so he won't be a king any more.'

Whatever the cause of King Lion's change of heart, they all quickly agreed that his roars sounded like a call to join him.

Serval proposed they should all go together at once to King Lion's hideout.

However, Jackal gazed around at them with his shrewd yellow eyes and said, 'Not so fast! We mustn't offend his majesty by all crowding into his court at once. It will be best to go there one at a time. Who would like the honour of being first?'

None of them had ever been near King Lion before, so no one volunteered.

'If everyone's too nervous,' said Jackal, 'we'll have to draw lots.'

He set to work organizing this, standing back politely to allow the others to take a turn.

Gerbil drew the winning lot. She twitched her whiskers bravely, blinked her large black eyes and set off without delay.

What a long journey it was for one so small! Moreover, being weak from dehydration, she could only travel very slowly. The other animals keenly watched her progress across the plain … lost sight of her at the foot of the mountain … then cheered when she came back into view as a tiny dot, now much higher up the slope.

Then suddenly, for the first time in any of their lives, they all saw King Lion. There he was – far, far away, standing on a ledge just below the highest peak. Even Jackal was awe-struck. King Lion turned round, melted into the shadows … then reappeared in front of a dark gap in the mountainside, right at the top.

'That must be the entrance to his den,' grunted Baboon.

And there was Gerbil too, still slowly approaching the king, crawling in and out of the crags. At last she arrived, her tiny frame silhouetted against King Lion's huge one … Before they vanished together into the darkness of his den.

Shrew said, 'Hopefully, once she's introduced herself and told the king about our plight, he'll send her back to fetch the rest of us.'

The animals waited eagerly.

Nothing happened.

'She's probably stuffed herself so full of King Lion's food and drink that she's fallen fast asleep,' said Hyena, licking his lips.

More time passed. King Lion emerged – alone – and roared again.

'You see, I was right,' said Shrew. 'He's summoning the rest of us to join them.'

'One at a time,' Jackal reminded them.

Hyena chuckled excitedly. 'Never mind drawing lots: I'll go up there next.'

He bounded off. The others watched his progress, across the drought-stricken plain and up the mountain ... until he too entered King Lion's den.

The others waited. More time passed.

King Lion came out and roared again.

'You must agree,' said Shrew, 'that's a very encouraging roar. I'm sure it means, "Hurry along, you're all welcome." What a wonderful time our friends must be having up there in the royal court! I volunteer to go next.'

So he too ran off and up the mountain.

After a long while, the roaring started up again.

In this way, one by one, the remaining animals all went to King Lion's den; and none of them came back.

At last, only Jackal was left.

Jackal walked across the plain and climbed the mountain much more cautiously than the others. When he was still a short way from the opening that marked the entrance to King Lion's den, he stopped and pricked up his long, keen ears.

Could he hear the cheerful voices of his friends drifting out from the interior? No. The place was completely silent.

He sidled a little closer.

A monstrously large shape appeared in the opening, framed by a circle of long, tawny fur.

'Good evening, King Lion, sir,' Jackal called out. 'I've come to check that all is well up here.'

King Lion yawned, flashing his teeth. Then he growled enticingly, 'Welcome, friend Jackal! What took you so long? Your companions are already feasting inside. Come in and join us!'

Jackal did not answer. Instead, he stood there sniffing, carefully examining the ground around the entrance to the den.

'Don't be diffident,' King Lion coaxed him. 'There's still plenty of tasty food left inside for you to share; and best of all, I have unlimited water.'

'Hmm,' said Jackal. 'Thank you for the generous invitation, your majesty – but I must decline it. There's something rather too familiar about the scent of fresh meat on your breath. And I can see the footprints of my friends in the dust out here, all leading into your den – but not a single one coming out of it.'

'Oh, don't worry your handsome little head about that,' roared King Lion. 'It's because ...'

But Jackal did not wait to hear his excuses. 'You're not eating *me*!' he shrieked. And he leaped back, spun round and raced down the hill, back to the plain.

There he resolved to fend for himself until the drought finally ended.

For, just as the soothsayer had advised King Lion, he knew that he must keep a clear head, and use plenty of cunning, to survive this terrible time.

Famine

Ireland

One year, shortly before the potatoes were due to be harvested, they were all infected with blight. The leaves became covered in brown and yellow spots then withered away. When the tubers were dug up, they were found to be rotten: brown inside, soggy, and completely inedible. The entire crop was wasted.

All the peasants depended on potatoes; it was their staple food. Since there were none left to harvest, everyone slowly starved. The same thing happened the following year. Those who could, fled the country, leaving the ones who remained to suffer slow, agonizing deaths.

The victims of this Great Famine soon started to apportion blame. Was it simply a natural disaster, an 'act of God'? Probably not. Some accused their heartless absentee landlords, others condemned their rulers.

But there were also people who attributed the famine to disquieting, unnatural causes. For in those days various kinds of uncanny folk still walked abroad, practising malicious magic.

A couple of years before the famine struck, as the dark nights of winter drew in, several people reported seeing a weird stranger lurking around the scattered cottages. The Hungry Man, they called him, because he was as thin as a rake, with hollow cheeks and tatty wisps of hair. He was dressed entirely in rags. Rumours said he was the ghost of someone who had died in wretchedness from lack of food, that he was resentful and full of hatred. Everyone dreaded him knocking at their door.

Was it just foolish superstition?

Surely not, because one family claimed to have actually had a frightening encounter with him.

They said that because of the gossip, they had already discussed what to do, just in case he came. So they were well prepared on the bleak, freezing winter night when they heard a feeble *rat-a-tat-tat* on the door.

'It must be him,' they whispered. 'The Hungry Man!'

The children were sent to hide under the bedcovers, pretending to be asleep. Then the husband let him in, while the wife bustled around, setting the table with all the meagre food they had left in their store. There was not very much of it.

'Come in, sir, make yourself at home,' she said nervously.

The Hungry Man tottered through the door. He was even more grotesque than they had imagined. His face was like a skull with shreds of skin stuck on it. His eyes had sunk so far back into their bony sockets that nothing could be seen of them except murky pinpricks. His hair resembled grubby thistledown, which also clung to his threadbare breeches and jacket. His feet were bare and calloused, covered in sores and cuts.

The husband pulled out a chair. The Hungry Man threw himself onto it, reaching out to grab the half loaf and small hunk of cheese that the wife had set upon the table. He shovelled them into his mouth, chomping noisily until he had swallowed the lot. That was the end of almost all the food the family had left until pay-day the following week; but neither wife nor husband dared stop their guest from sating himself. Indeed, after he had licked the plate clean, the husband nudged the wife, who said hastily,

'I hope you've eaten your fill, sir? I'm afraid our cupboard's completely bare now, but there are still a couple of small cabbages growing in the yard, and you're welcome to those if you wish.'

She managed to control her shuddering and keep the smile fixed on her face.

The Hungry Man fixed his pinprick eyes on her, speaking in a voice like a hoarse, painful groan: 'Cabbages? No thank you, lady, you can keep them for yourselves. My belly will always feel empty, but you've been generous enough.' He rose unsteadily to his feet, shuffled to the door and opened it wide, letting in a rush of icy air. 'God bless you for your generosity, lady, and you too sir. You'll get your reward.' Then he vanished into the night.

Not long after this, the family arranged to emigrate to a better country. All their relations and friends came to see them off when they boarded the ship, and asked enviously how they had managed to find the price of the voyage. That's when they revealed their encounter with the Hungry Man. They said that only a short time afterwards, money came their way completely unexpectedly – proof that he had kept his promise.

Unfortunately, others weren't as wise as them.

For a long while, no one saw the Hungry Man again. By the following winter, talk of him had faded right away.

One night, another family of cottagers settled down in front of their fire, darning, spinning and mending tools. As the peats burned down to glowing embers, they heard a feeble *rat-a-tat-tat* at their door.

'Who's there?' the wife called out, expecting the familiar voice of one of her kin.

No one answered.

Rat-a-tat-tat.

'A stranger, eh?' the husband shouted. 'Get away with you at such a late hour, stop bothering us!'

'Hush,' the old grandmother scolded him. 'It might be someone in trouble, needing help.'

A quavering voice called out from beyond the door, 'Yes indeed, will you help me? I'm starving. Spare me a rind of bacon and a morsel of bread.'

'Certainly not, you good-for-nothing beggar!' the husband shouted back. 'Go and earn yourself an honest living, like the rest of us have to.'

The grandmother shook her head at him earnestly. 'Son, you shouldn't ...'

Outside, the stranger rattled the latch. The husband jumped up to turn the key in the lock. But before he could do so, the door burst wide open – and the skeletal Hungry Man staggered in.

By now, his remaining shreds of skin had almost withered away. His thistledown wisps of hair were tangled with green mould. His breeches were missing one leg, and the skin of his feet had worn away almost to bare bones.

The whole family gasped.

'I'll not forget your callousness tonight,' the Hungry Man hissed at them. 'You heartless brutes, you'll suffer for this! Your whole family will suffer, your whole village will suffer. The entire country will suffer! In this coming year, all your crops will fail. Your stomachs will churn and shrink, then distend with hunger. Paupers, that's what you'll become: starving nobodies like me. Meanwhile, the rich who rule this land will get even richer – but they'll show no pity for you, just as you showed none for me. Next year, your crops will fail again. That's what

I promise you in revenge for your lack of charity: Famine! Boundless suffering and early death.'

With that ominous curse, he slammed the door shut – and was never seen in the land again.

No one knows whether the Hungry Man was really to blame for the Great Famine, or indeed whether he even really existed.

But one thing is beyond doubt: everything he spoke of in his curse came true.

Litter

Colombia

There was a vast forest. In the middle of the forest there was a lake, and on the shore of the lake stood a village. This village had so much food readily to hand that the people who lived there hardly needed to do anything to stay well fed. They didn't bother growing crops; nor did they keep livestock or hunt animals for meat. Instead, they lived entirely on fruit.

The fruits dangled from countless wild trees that surrounded the village in all directions, and it was always the right time to pick at least some of them. There were over ten different kinds of mangos, alongside bananas, oranges and countless other delicious ones that outsiders haven't even heard of. Also, wherever the villagers walked through the forest, they kept almost tripping over luscious ripe melons, and trailing vines laden with juicy grapes. Not that anyone who lived there ever wandered very far. Whenever they felt hungry, they simply heaved themselves up from where they were sitting, walked languidly to the nearest tree, shifted their hands up just far enough to pluck a tasty fruit, then stuffed it into their mouths.

It was idyllic.

There was only one drawback: and that was the litter. Because all the non-edible bits of fruit – the rinds and peelings and pips and pith, the stalks and stones – just got dropped carelessly to the ground and left there. Why bother to clear them up? It was so much easier to leave them where they were to rot.

But even in a tropical climate, rotting takes several days, sometimes even longer. And until the process is complete, mouldering edible rubbish attracts swarms of pests – particularly insects.

At first, it was only butterflies that came to gorge themselves on the decaying juice. The lazy people weren't bothered at all by these gentle little creatures with their fluttering bright colours. They even threw down pieces of fruit deliberately to attract them.

But in their wake came others that were more like miniature monsters: flies, bugs, bees, mosquitoes and wasps – *zzzzzzzzzz!* – *thousands* of them. Whenever a human being got in their way, they reacted by biting and stinging viciously. *Zzzzzzzzz! Zzzzzzzzz!* Sometimes they made a direct attack without even bothering to feed on the waste. And if people tried to swat them off it just made them even more aggressive.

Before long, every single villager, young and old, was covered from head to toe in painful bites and wounds. Some victims fell seriously ill as a result.

Finally, the village headman stirred himself into action. He called everyone together to discuss what to do about the insect infestation. One by one the listless villagers shuffled up, then immediately yawned and sank back to the ground, scratching madly at their bites.

'We can't carry on like this,' said the headman. 'Can anyone suggest what to do?'

There was a long pause while they all blinked and fidgeted. It was hard work trying to force their brains into thinking. None of them were accustomed to such effort.

At last, a woman said slowly, 'I suggest we clear up the mess we've made. Then the insects won't have any reason to bother us any more. And from now on, we should stop throwing our rubbish around and make an effort to keep the place tidy. That will stop the pests from coming back.'

'Clear up, did you say?' scoffed the man opposite her, letting out a slow burp. 'You must be joking! It's far too much effort. There must be an easier solution.'

'Think how dirty it would be, touching all that rotting fruit,' said another woman. 'Ugh!'

Despite the first woman's nagging, everyone else agreed that removing their litter was not something even worth considering.

All through the languid afternoon, other suggestions were made – and immediately dismissed as being either too strenuous or too unpleasant. The headman himself sat there silently, deep in thought. At length, he clapped his hands for attention and said, 'My friends, I have the perfect solution. This forest is endless, and the lake which provides our water is huge. We don't need to stay here. We can simply leave our rubbish behind and move to a brand-new location along the lakeshore, right away from the mess and the insects. We can build some new huts in a nice clean place there. In this way, we'll leave our problem behind.'

'Yes!' everyone cheered. 'How lucky we are to have such a wise leader!'

'But won't it be hard work building new huts?' said another man in a worried voice.

'Not as hard as clearing up our litter,' the headman assured him.

'At least it's clean work,' said the second woman.

Thus, it was agreed, and so it was done. The people didn't really have any possessions, so all they had to do was gather their children together, then walk through the forest. Naturally, they went at a very slow pace, singing and picking fruit to munch as they went.

By the end of the day, they had reached a pristine new spot on the lakeshore. 'We'll make our new settlement here,' the headman announced.

Before starting work, they sat down and feasted on the fruit of the surrounding trees, none of which had ever been touched by human hands before. After that, the women gathered more armfuls of fruit to lay aside for their breakfast, while the men stirred themselves to cut down some bamboo and weave it into simple shelters.

By the time they had finished these tasks, on top of their trek through the forest, they were more exhausted than any of them could remember. So they kicked the remains of their arrival feast carelessly out of the way, then crawled into the new huts and fell fast asleep.

The next morning, they all got up late and started on their next meal.

'Nice and clean here, isn't it?' said a plump father of many children, as he stuffed himself with grapes, then spat out a mouthful of pips.

'Wonderful to be free of insects,' his wife agreed, tossing banana peel over her shoulder.

They all settled comfortably into their new homes. For a blissful few days, they heard not a single disturbing sound, and the itchy nips and punctures on their skin completely healed.

But what do you think? Yes, before long, those lazy people had created fresh heaps of rotting litter all round their new village. And soon they heard that familiar, unwelcome sound again: *zzzzzzzzzz!* For the insect colonies had found them! They arrived *en masse*, attracted by the disgusting mouldering smells. The lazy people were again plagued by a torrent of painful stings and bites, just like before.

This time, as they swatted the vermin away, the headman knew exactly what to do.

'Heave yourselves up off your backsides!' he ordered. 'It's time to move again.'

So once more, they got up and trundled through the forest along the side of the lake.

They found a fine new spot, as completely untouched as the previous one, and full of the usual bounteous fruit trees. Again, they erected some ramshackle huts, then settled down to spend their idle days feasting on the free forest fruits.

What a wonderful way of life! Now they had found the solution of moving on whenever the insects caught up with them, they expected it to last forever. And indeed, it went on for many months: settling somewhere new for a while, filling a formerly spotless place with litter, then abandoning it to move on even further.

To ensure they didn't run out of water, they always kept to the lakeside. Which meant that by the time a whole year had passed, they had travelled in a full circle – all the way around the shore – and arrived back at the very place from where they had started!

The first they knew of it was a nauseous stink, wafting towards them as they drew near. Then they spotted familiar mounds dotted around the ruins of their old huts, festering with mould and maggots, surrounded by droning, fizzing clouds of tiny, winged beasts.

For their original heaps of litter were still exactly where they had left them. And there were now similar heaps at intervals all the way round the lakeside.

There was no escape from their mess.

Polluting the Sea

Netherlands

In a splendid city by the sea, a widowed lady lived alone in great comfort and splendour. Her late husband had been a prominent merchant, and when he died, his entire fleet of ships passed into her keeping. They were constantly arriving in port with the most valuable kinds of cargo, bringing her ever-increasing profits. Thus, her mansion overflowed with gold and gemstones, antiques, porcelain, paintings, statuettes, tapestries, furs, silks and countless other precious things.

But just like that old saying, 'Much wants more', this lady was always dissatisfied and yearned for even greater wealth. So one day she announced a contest to bring her 'the Most Precious Thing in the world'. If any man could identify what this was, and present it to her in perfect condition, she promised to take him as her second husband.

All the unmarried men of the city were greatly intrigued by this idea. For the lady was still youthful, and as beautiful as she was rich. There was much coming and going, with potential suitors travelling all over the seven seas to bring her the most extraordinary treasures. However, though she readily accepted their gifts, she quickly rejected all the men. For she complained that none had yet brought her the elusive, undefinable – perhaps impossible – Most Precious Thing that she wanted.

Eventually, a sea captain with his own large merchant ship thought he had solved the riddle. He was an unpretentious, down-to-earth man who had achieved success by straightforward hard work. He sent the lady a courteous, beautifully scripted message saying that he was about to sail on a quest to bring back her heart's desire.

By this time, the lady was weary of being bombarded by swaggering men who all fell far below her expectations. This captain sounded well-bred and dignified, and she liked the low-key tone of his message. So she sent back word to say that she eagerly awaited his return.

The captain was away for many months, visiting numerous foreign ports. In each one he took on board a large amount of secret cargo, loading it discreetly inside tightly sealed barrels. By the time he arrived back at the seaport where the lady lived, the place was brimming with rumours and gossip. He went ashore and strode past the prying crowds, straight to the lady's mansion.

The lady received him graciously in her grand salon, relishing his good looks and well-cut clothes. 'Well,' said she, 'have you managed to bring me the Most Precious Thing in the world?'

The sea captain kissed her proffered hand and replied, 'Madam, I have indeed. It is too big for me to carry up here, but may I invite you down to view it on my vessel?'

'First tell me what it is,' she ordered.

The sea captain bowed and said, 'Madam, I have brought a ship crammed full of the world's finest wheat. It is all yours.'

The lady stared at him incredulously and snatched away her hand. 'Did you say *wheat*?'

The sea captain smiled and nodded.

'What nonsense is this?' she exclaimed.

'Nonsense?' said the captain. 'Of course it isn't. Wheat is the very essence of life itself. No one can live without their daily bread; and bread cannot be made without its main ingredient, wheat – which is thus, beyond doubt, the Most Precious Thing in the whole world.'

The lady had turned white with rage. '*I* can live perfectly well without bread, you ignorant wretch! That's only fit for paupers and beggars. All *I* need is treasure. How dare you offer me something so lowly instead!'

The captain shrugged. 'We clearly have different ideas about what is valuable, madam. My cargo is enough to feed this entire city for a whole year. No one can survive by eating exotic things like gold or diamonds. Well, if you can't make use of the wheat yourself, please donate it to your many neighbours who are undernourished or even starving.'

'Never! You brought this gift for me. I'm not giving it away to good-for-nothings.'

The captain shrugged. 'Rather than offend you any further, madam,' he said quietly, 'I withdraw my offer to give the wheat to you. Instead, I shall donate it to our impoverished fellow citizens, who no doubt will truly appreciate it.'

'Now you're adding insult to injury!' she cried. 'You have already said this useless cargo is mine – which means I can use it however I wish. Well then: I shall have the whole lot thrown away.'

'No, madam!'

'Yes!' she retorted. 'Go back to your ship, captain, open your storage hatches – and tip every single despicable grain of wheat straight into the sea.'

'But that's total wastefulness!' said the captain heatedly. 'It's like snatching bread from the mouths of people who need it. Besides, there's so much, that it will probably block up the harbour.'

'That's not my problem,' she said coldly. 'Do what I say, captain. This so called 'gift' of yours is nothing but rubbish. Dispose of it accordingly.'

The lady refused to listen to his arguments and appeals, and her power was far greater than his. To ensure the deed was done, she ordered her private bodyguards to manhandle him back to the ship. Then she set them to supervise the discharge of his entire load of wheat. Out it poured, like a torrent of brown hailstones, straight into the previously sparkling harbour water.

For a long while the grains floated on the surface, a mass of flotsam and jetsam. The people of the city watched them in dismay, but the grains were too numerous and too small to be removed. Gradually, the tides carried them all away to the harbour mouth. There they swelled up, mingled with the drifting currents of loose sand, then sank.

Months passed.

By the following spring, the wheat grains had germinated, taken root, sprouted and grown into a mass of tattered looking weeds. As they matured and became taller, more and more swathes of floating sand got caught up in them. In this way, they created a wide sandbar.

It completely blocked the harbour entrance. As a result, the merchant ships could no longer come and go.

Above the clogged-up water, the harbour became derelict. And the once grand city fell into decline.

Harming Animals

Arabia

It is said that in days of old and times long past, there lived a young lion who had reached the age to venture out into the world alone. So his wise old father called him aside and spoke to him as follows:

'Listen carefully, my son. You already know that we lions are far mightier and fiercer than any other animal. However, I must warn you to be constantly on your guard against one species in particular. At first sight, they appear to be relatively feeble. But the truth is that they are masters of insidious, deadly cunning.'

'What name do these creatures go by, father?' the young one asked.

'Humans,' his father replied. And as he spoke this unsavoury word, a shudder ran through his sleek, tawny body. 'If ever you come across one, be on your guard!'

'But how would I recognize it, father?'

'Oh, that is easy,' said Old Lion. 'Humans have only two legs, no tail and so little fur that they must wrap their naked bodies in cloth. Keep a careful lookout for them, my son. If ever a human approaches you, immediately pounce and kill it. For they play no useful role in the world; they are nothing but dangerous vermin.'

The pair rubbed their heads together affectionately. Then Young Lion thanked his father for the excellent advice, and set out to explore.

He walked a long way until at last, in the middle of the desert, he came to a crossroads. He stood gazing at it for a long time. Whatever kind of marvellous being had enough skill to smooth the sand into four separate, perfectly straight tracks like this, each stretching away to the distance?

As he pondered this, he suddenly saw a cloud of dust swirling up above one of these tracks, and heard the thundering of feet. The dust faded away to reveal an unknown type of creature, which halted abruptly in front of him. Young Lion hastily examined the newcomer, ready to attack if it turned out to be one of the dreaded humans. Fortunately, however, it was a respectable animal, completely covered in short fur. It had the usual four legs, a tail quite similar to his own, and even a short mane running down its neck.

'Who are you?' Young Lion asked politely. 'And where are you running to, so fast?'

'My name is Ass,' the other replied. 'And I am running away from a human.'

Young Lion's heart missed a beat. 'A *human*?' he whispered. 'How terrible!'

'Yes, indeed,' said Ass. 'If this human manages to catch me, he will immediately start abusing and torturing me – as he has done all my life, for I am his slave. He'll beat me with a spiked stick until he's forced me to run twice as fast as I can comfortably manage. If I stumble or slow down to catch my breath, he'll beat me even harder.'

'Then just as my father told me, humans really are evil!' said Young Lion in horror.

Ass gave a wretched laugh. 'You're right there! Misery and cruelty – that's my lifelong fate, young one! Let me pass now, if you will. No doubt the human I'm fleeing has already discovered that I'm missing and is coming after me. I must urgently find a place to hide.'

Young Lion stood aside, and Ass galloped off, veering away from the road into the wilderness. The next moment, another dust cloud appeared over a second track. Very soon a larger creature emerged from it and rushed up to the crossroads. This one was very black and very handsome – truly magnificent. It also had four legs and neat fur, with a thick tail, and a mane almost as splendid as Young Lion's father's.

'Who are you?' Young Lion greeted it. 'And where are you running to so fast?'

'My name is Horse,' the other replied breathlessly. 'I am trying to escape a human.'

Young Lion looked at the newcomer in surprise. 'But I've been told that humans are very feeble – so how can they hurt a well-built animal like you? Surely with your strong legs, you could easily get away from such a pest – or simply kick it to death?'

Horse whinnied mournfully. 'So you think, young one. But humans are so crafty, they can easily trap my kind, no matter how strong we are. I was taken captive as a young foal. My human master hobbled my legs with rough cords. Then he roped my neck to an overhanging tree branch, so it was impossible to walk, sit or lie down. *Neigh-ch!* It was slow torture! When the human had completely broken my spirit, he forced a strip of bitter iron into my mouth, attached to straps all over my face. Then with

more straps he fixed a bulky leather saddle on my back and climbed onto it – the clumsy, heavy oaf! He used his feet to dig sharp spurs into my flanks until they bled. I reared up and tried to throw him off – but that only made him spur me even harder, and whip me mercilessly until I moved the way he wanted to go. It has been the same every day since then. I have nothing to look forward to until my throat is eventually cut by the knacker, who will flay my skin to sell for leather, pluck my beautiful tail to be made into sieves, and melt my fat down for candles.'

'How terrible!' Young Lion exclaimed.

But Horse had already raced away into the wilderness – and another dust cloud was swirling above the third track leading to the crossroads.

The creature that came along this one was even bigger than Horse, and rather strange looking. Its four long, elegant legs carried a massive body, its neck was oddly curved, and its back was topped by a rounded hump.

'Help me, young friend,' it grunted. 'A human's after me!'

'Who are you?' asked Young Lion.

'My name is Camel,' the other replied.

'But you look so sturdy,' said Young Lion. 'How can *you* be afraid of a human?'

Camel snorted. 'Ach, how ignorant you must be to ask that! Look at me properly. Can't you see the nose ring that this human shoved into my nostrils, with the loose rope trailing down? He normally uses that to drag me along behind him, after overloading my back with heavy packages. I suffer a life of endless hard labour. If I try to resist, the human threatens to kill me and cook me. Move out of the way, young one, so I can flee before he catches me up!'

By now, Young Lion was determined to help these poor creatures against their oppressors. He was almost fully grown, and certainly fierce and strong enough to do so. So instead of moving, he said, 'Fear no more, friend Camel. When that human comes, I shall tear it to pieces! I'll serve you its flesh to eat, while I crunch its bones and drink its blood!'

'It's easy to say that, young one,' Camel said gloomily. 'But you have no idea what humans are really like.'

He pushed Young Lion gently to the side and ran off.

Shortly after that, Young Lion saw a fourth dust cloud, and heard the drumming of feet that announced the approach of yet another victim.

This time, he was determined to find out exactly where these cruel humans were lurking, so he could kill the lot of them!

Soon the new animal was standing before him. It had rather a different physique from the others. For it stood upright on just two legs, and had no tail. Its body was hairless and covered with a large cloth.

Young Lion had not forgotten his father's sombre warning. However, it could not possibly apply to *this* sorry creature, which seemed in even more distress than the other three. For it was staggering under a terrible burden: six heavy planks of wood balanced precariously on its head, and two overfilled baskets weighing down its shoulders.

'Oh, you poor creature!' Young Lion exclaimed. 'Who are you?'

'My name is Carpenter,' was the reply. 'Step out of my path, young one, for I must hurry on my way.'

'What for?' asked Young Lion.

Carpenter gave him an unsettling look. Then he cleared his throat and said, 'To build a sanctuary for the poor animals who are constantly tormented by humans.'

'Oh! How generous and helpful you are, friend Carpenter!' cried Young Lion. 'Three of these suffering animals have only recently passed this way. They're all desperate for a safe place to hide and will be delighted to hear that you'll build one for them.'

Carpenter threw him a kindly smile. 'Tell me which way they went, young one, so I can quickly catch them up.'

Young Lion considered this for a long moment. He recalled that neither Ass, Horse nor Camel wanted *anyone* to know where they had gone. Besides, he wasn't naïve, oh no, not him; and he had heard so many frightening tales, that he himself now longed for a sanctuary. So he said, 'Don't worry about helping the other animals, Carpenter. Just build *me* a hiding place. I have the greatest need of all to keep safe from these dangerous humans.'

Carpenter narrowed his eyes at Young Lion. 'Has any human ever directly hurt or harassed you?' he asked.

'No,' Young Lion admitted.

'Then I expect you've never actually met a human – am I right?'

'Yes,' said Young Lion.

Carpenter chuckled to himself, then looked at Young Lion severely. 'You must wait until I've helped those who have already suffered from human wickedness,' he said. 'Tell me which way they went. Once I've attended to them, I'll come straight back to help you.'

But Young Lion had not been brought up to await his turn. 'No, Carpenter,' he said stubbornly. 'Build *me* a shelter – now!'

He drew himself up to his full size, bared his fangs, let out a fierce roar – then charged at Carpenter, striking him hard with his massive paw. Carpenter went sprawling to the ground, scattering his planks and tools.

Young Lion roared again: 'Build me a shelter at once, or I'll tear you to pieces!'

Carpenter staggered back to his feet with an unreadable expression on his face. 'Don't kill me, powerful one!' he begged. 'All right, I'll do exactly as you ask. I'll make your shelter here and now – a very special one, exactly the right fit for you.'

'Good,' said Young Lion gleefully.

Carpenter set to work. He gathered up the planks and pulled various tools from his baskets. He told Young Lion to sit still, then set to measuring, sawing and hammering. In no time, he had fashioned five sides of a cube-shaped hut.

Young Lion paced around the outside, admiring it from every angle. 'It's certainly a perfect size for me,' he said. 'But there's just one problem. The front is wide open. If a human finds it, he could easily throw spears into it, or drag me out.'

'How clever you are to point this out,' said Carpenter.

Young Lion shook his mane and grunted with pleasure at this praise.

Carpenter went on, 'In order to make the shelter secure, I need you to step inside it.'

'Ah,' said Young Lion, 'is that to fit a door?'

Carpenter said nothing, but ushered him into the box.

'It's a bit small,' complained Young Lion. 'I can't stand up properly.'

'Oh, you don't need to,' said Carpenter amicably. 'Lie down, young friend, and make yourself comfortable – knowing that you're perfectly safe in there.'

'But I can't turn round in it.'

'You don't need to do that either.'

So Young Lion squeezed right in. 'My tail's still outside,' he called.

'It isn't now,' said Carpenter.

He stuffed Young Lion's tail into the box, so roughly that the animal let out a yelp of pain. Carpenter ignored it, rammed the last plank over the opening, and quickly nailed it in place right round the edges.

'How do I open the door?' came Young Lion's muffled voice from inside. 'Where's the catch?'

Carpenter laughed. 'The 'catch' is that you *can't* open it.'

'But I thought ...' cried Young Lion.

'You thought *you* were the clever and powerful one, didn't you?' Carpenter called back. 'Yet you have fallen straight into my trap. Don't you realize? Though I use the pseudonym Carpenter, I am actually a human. Yes, one of those that all the other animals dread. And now you have discovered for yourself the truth of what everyone says: No other creature in the whole wide world, no matter how strong and fierce they are, can ever outwit human cunning!'

War

Sioux people (Native American and Canadian First Nation)

In the early days, human beings did not dominate the world, for two groups were far stronger than us. Unfortunately, there was constant conflict between them. This had terrible repercussions for the people and animals who lived at that time, and also for the Earth itself.

One of these groups, the Water Monsters, were manifestations of pure evil. They looked like huge, scaly snakes with feet, and had long horns protruding from their foreheads. Their chief was a female. She lived at the bottom of a river, filling its bends and curves all the way from one end to the other. Sometimes when the whim took her, she liked to puff up her body, making herself really fat. This made all the water in the river around her overflow, deluging the land. Her numerous offspring would then copy her in the smaller streams and lakes which they inhabited, until everywhere was overwhelmed by floods.

The people didn't complain or try to oppose the Water Monsters. They had learned from bitter experience that it was impossible to overcome them by any means. The Water Monsters despised what they saw as human weakness. 'Pathetic, dirty lice,' they used to call us, and 'annoying little blood clots.'

The other powerful group, the Thunderbirds, were even more formidable than the Water Monsters. They had extraordinary supernatural skills, controlling the sky and making storms. Their four male chiefs, who all had many children, wore robes of dark clouds and lived on top of a high mountain. Their domain could only be accessed via four secret paths, which were variously guarded by a butterfly, a bear, a beaver and a deer.

The Thunderbirds were a force for good, and guardians of truth. They were benevolent towards human beings, whom they treated with respect and care, like distant kin. They very much appreciated it when the people offered up prayers to them, and often responded by appearing in dreams and sharing fragments of their power.

One day, the Water Monsters flooded the land particularly badly. They not only puffed themselves up, but also splashed around so violently that the water rose higher than ever before. Most of the wild creatures drowned. All the people's homes were destroyed and many of them drowned too. The survivors had no choice but to climb to the top of the highest mountain and wait until the flood subsided.

But it didn't. This isn't like other stories about cosmic floods that you might have heard. No matter how long the people sat it out up there, the Water Monsters kept swelling up again, spattering great waves all over the place, leaving the people's homeland permanently submerged. This enabled the Water Monsters to control all the low-lying regions of the world, without being challenged.

So the Thunderbirds declared open war on them.

They made this declaration in a mighty clap of thunder, echoing loud enough to be heard right across the world. This was followed up by their children's ominous rumbles and booms. The enemy was left in no doubt of their intentions.

At first the war was quite conventional. The Thunderbirds swooped down on the flooded waters and attacked the Water Monsters. They tore at them with their huge, knife-like claws; they bit them with their beaks and razor-sharp teeth, scattering monstrous blood everywhere.

Unfortunately, however, this quickly became mingled with an equal amount of Thunderbird blood. For the Water Monsters thrashed back with their heavily spiked tails, gouging terrible wounds in the Thunderbirds, even damaging their all-seeing eyes. Because the battle was taking place in the Water Monsters' own wet element, they easily gained the advantage.

So the Thunderbirds withdrew to rethink their tactics. Back on their mountain top, their chief gathered them together in council.

'My friends,' he said, we have no hope of defeating our enemy if we only fight the battle down on the ground and in the water, for the Monsters have intrinsic power there. Since our own strength and inspiration, comes from the sky, it is from here that we must fight our next – our *final* – battle. I greatly regret that we must use unrestricted violence. But this is the only way to save the world, so that human beings and their fellow creatures can eventually recover and flourish.'

So it was agreed. The Thunderbirds launched an assault of thunderbolts and lightning from the sky, directly onto the Water

Monsters. Such an attack had never taken place before – and it was devastating.

The flooded waters heated up to boiling point, bubbling away as furiously as cooking pots on blazing fires, until they had completely dried up. This completely destroyed the Water Monsters' domain. The grassland and forests all caught alight and burned for months until nothing was left except blackened stubble. The whole world got so hot that it glowed bright red.

This conflagration certainly finished off the Water Monsters. The lightning set fire to their evil chief and her equally wicked offspring and cronies, burning them all to cinders.

All through the fighting, the Thunderbirds took great care not to harm the handful of people still clinging to life up in the mountains. Eventually they were able to come down and remake their lives on the plains. Some of the animals also survived – just enough to fulfil the people's needs for food, shelter, tools and companionship.

Gradually, the Earth was repopulated by both human beings and animals. But it took hundreds of years to return to normal after all the flooding, burning and other destruction.

According to the Thunderbirds, this devastating war was the only possible way to destroy the cataclysmic evil that threatened the Earth's very existence. Ever since then, many human leaders have made similar claims.

Whether or not such claims are true, wars are always unquestionably catastrophic for the natural world.

Extinction

Nigeria

There was once an extraordinary hunter whose arrows and spears never missed their target.

Such a skill does not come easily to a man. He has to carefully follow all the correct rituals before he sets out: bathing, oiling and painting his body; abstaining from certain foods and the company of women; dancing to make himself invisible to his prey. The hunter did all this diligently. Even better, he developed a curious way of whistling that never failed to attract the animals he was after.

Quite early on in his career, working alone, he killed ten large animals in a single day. Everyone was very impressed, especially as he regularly repeated this tally. Then he started bringing home 20 animals, 30 – even 50, all yielding top quality meat and hides.

Suspicion spread that he must somehow be cheating. Perhaps he employed a team of elusive assistants to hunt alongside him, or had supernatural powers? So a group of men spied on him. Reluctantly they reported back that everything was exactly as he claimed. They had witnessed with their own eyes how exceptionally talented he was at stalking and shooting game, never once failing.

Soon after that, his daily kill increased even more, sometimes up to a hundred animals. Yes, a hundred in a single day!

Of course, that was far more than he needed for himself. So he sold most of the game he brought back, making him very rich. He came to own fields full of cattle, alongside goats and pedigree horses. All the district chiefs courted his favours.

But though he no longer needed to hunt, he still continued to do so every day. No matter that he had cleared out so much game that all his rivals kept coming home shamefully empty-handed. No matter that now even he had to trek much further into distant reaches of the forest to find his increasingly rare, rapidly diminishing prey. It was a compulsion, an addiction: the hunter simply could not live without his daily multiple kill.

After this had been going on for a long time, the hunter had a disturbing experience. He had killed his usual haul, and arranged the carcasses in neat piles, ready for his men to drag back to the storehouse. As he ambled cheerfully home on his own, he suddenly heard something that brought him up short. A curious, reedy voice was calling out:

> *You!*
> *Stop where you are!*
> *Listen!*

The eerie tone sent shivers down his spine. He looked around to see who was there: no sign of anyone. He shifted his spear, ready to throw at a potential assailant, then continued walking.

> *You!*
> *Did you not hear me?*
> *Stop!*

Where was it coming from? He checked every direction, but nothing appeared out of place. There were no unexplained shadows or movements among the thick creepers and trees.

'Who are you?' he called back. 'Stop hiding from me, coward. Show yourself!'

Nobody appeared. But a sinister presence seemed to emanate from the undergrowth and envelop him.

> *You cannot see me,* said the voice, *but you most certainly know me – as well as I know you. I am the god of this forest.*

For a long moment, it paused. Everything was completely still.

A lesser man might have shuddered; but the hunter gave a confident laugh, slapped his thigh and strode on.

The voice continued:

> *I have watched you. Yes, through many long days I have observed your misdeeds, your reckless pride in slaughtering unlimited animals. I have seen you amass extraordinary wealth by selling your kill. You are proud*

of the countless cattle you own, are you not? You relish enabling your adoring family to feast on your riches. You gloat over the admiration of your fellow men. But in devoting your life to untempered slaughter, you neglect a far more vital skill: that of allowing enough wild creatures to survive and multiply for the future. Learn to do this now. Start controlling your greed at once. Otherwise soon there will be no animals left to feed your people, and they will die.
Listen and obey!

By now, the hunter had indeed stopped and was listening intently. But everything fell silent, and again he seemed to be alone.

He resumed his walk home, wondering if he had imagined the whole experience. True, he had heard elders talking of dangerous spirits that prowled the forest, but no one he knew had ever actually met one. Most likely, it was just his mind playing tricks after another long, exhausting day. *Control your greed,* it had ordered him. What arrogant nonsense! There was nothing greedy about supplying everyone with plentiful meat! And it was only right for him to be applauded and paid well for his exceptional efforts. Ordering a hunter to stop killing was total nonsense.

The next day he set out as usual for the area where game was still fairly plentiful. He stalked his way softly through the trees there, and was soon rewarded by the sight of three huge antelopes grazing docilely in a clearing. He grinned to himself, strung his bow and shot them quickly, one after another.

As the last one fell to the ground, the disembodied voice started up again:

So: you ignore your duty to preserve the creatures that sustain you. You shall pay for this neglect with your life!

He had no time to respond. For his eyes were riveted by the dead antelopes. All three had suddenly started to shimmer, like sunlight on dappled water. A great coil of mist rose from the ground where they lay, swallowing them up. Gradually, very slowly, the mist folded in on itself … and became a solid shape … which separated into three forms … no longer antelopes – but *lions.*

The lions threw back their heads, roared – and sprang at him.

But before they could touch him, the hunter had already drawn on his own secret powers. He too shape-shifted – into a small bird – and quickly flew away.

The lions' fangs snapped at empty air, and their huge paws struck at nothing. But at once they transformed again – this time into hawks. They spread their powerful wings, soared up into the air and surrounded the little bird …

Which at once changed into an enormous tree, solidly rooted to the forest floor. The hawks were flying so fast, surely they could not avoid crashing into it and stunning themselves …

But just in time, they transformed into fire.

The flames took hold of the tree before the hunter embodied in it had time to shape-shift again. Within moments that rapacious destroyer of species was burned to death, with the fire's anger smouldering around him.

For fire once fire gathers momentum, it is almost unstoppable.

It is exactly the same with extinction.

RESPECT THE EARTH!

The Earth is Our Mother

Nespelem people (Native American)

Old One made the Earth out of a woman,
and said she would be the mother of all the people.
Thus the Earth was once a human being,
and she is still alive.
But she has been transformed,
and we cannot see her in the same way
that we can see a person.

Nevertheless, the soil is her flesh,
the trees and vegetation are her hair,
the rocks are her bones,
and the wind is her breath.
She lies spread out, and we live on her.

Beware of Consequences

Myanmar

A king was once eating his breakfast when a few drops of honey fell from his spoon to the floor.

Of course, being a king meant he was far too important to clear up his own mess. Besides, it was so insignificant, why worry about it? So he carried on eating, leaving the spilt honey where it was for the servants to attend to. He finished his breakfast at a leisurely pace, belched loudly, then strode off to attend to matters of state.

The spilt honey was spotted by a fly. It hummed its way down and greedily sipped it.

Lurking nearby was a lizard. It darted forward and snatched up the fly on its long tongue.

This quick movement attracted the notice of a palace cat, which was half-dozing in the corner. It pounced on the lizard and started tossing it around in its mouth like a toy until the poor creature died of fright.

The cat's callous game caused such a commotion that one of the palace dogs ran in through the open door to investigate. Seeing its archenemy the cat, the dog started to chase it. The cat fled, racing round the breakfast chamber, out of the door and into the garden, with the dog at its heels.

The dog's excited barking made the servants rush out to see what was going on. The man who fed the cat each day yelled at the dog's master to call his pet off at once. The master tried, but the dog took no notice. Instead, it leaped on top of the cat and bit it to death.

The man who fed the cat was outraged. He punched the dog's master and wrestled him to the ground. Then he drew out his sword and killed him.

The other servants had been standing around, watching. They now all took sides to form two opposing gangs.

The dead man's friends rushed to attack the killer. But the killer's own allies flew into the affray to defend him. Within moments, it was a full-scale battle.

There was so much noise that the king himself came back in to see what was going on. He strode towards the fight, shouting at the men to be quiet, ease off and split up. But both gangs ignored the king's order, told him to mind his own business – then started attacking him too!

His guards came running in, but it was too late – for the king was already dead.

The guards separated the two violent gangs and called the senior councillors to interrogate them. Each gang accused the other of having set up the situation in advance, as an excuse to kill the king and seize power for themselves. It was impossible to discover the truth.

The enmity spread throughout the land and developed into a full-blown civil war. By the time it was finally resolved, the whole country was strewn with broken weapons and the filthy rubble of exploded buildings.

And to think that it all started with a few innocuous drops of spilt honey!

If only that king had anticipated the trouble he would cause by not bothering to clean up his own corner of the environment! It wouldn't even have taken much effort. Then he could have saved not just himself, but also his entire realm, from defilement and destruction.

Value Everything

Maya people (Central America)

There was once a labourer who had always worked for other people – which meant that he hardly earned any money. He was desperate to run his own farm. He dreamed of growing enough crops there to feed his family well, and also to sell the surplus. In this way, he hoped to throw off his wretched poverty for good.

Unfortunately, he had no savings, so he couldn't afford to buy any land. But he lived near a wide stretch of untouched forest. One day, when he was exploring it with a bag of tools on his back, he found a small clearing that would be perfect for his needs. The only problem was that right in the middle of this clearing stood a large tree.

Well, a single tree was easily dealt with. He regarded it for a few moments, admiring its bright orange flowers, then raised his axe to chop it down.

Before he could do so, the air was suddenly rent by an uncanny, heart-rending shriek:

Stop! Don't destroy me!

The farmer dropped his axe in astonishment and gazed around. Who could it be? What was going on? He had deliberately chosen a remote place to avoid conflict with other people, and he had been careful to check there was no one around to challenge him.

'Where are you?' he called. 'Show yourself!'

Nothing moved. He heard no sound: no footsteps, no rustling, no twig cracking.

He must be imagining things. But then he heard the voice again:

Right in front of you!

He blinked. Still, there was definitely nobody there except the tree. He picked up the axe again, swung it back and …

Human, don't strike me! Let me live, and I will shower you with treasures.

It was so eerie that the farmer began to tremble. 'Who ... who is speaking?' he stammered.

It is me: Ciricote Tree.

'A tree? Don't try to fool me, trees can't talk!'

In desperate times, we find our voices. And I am desperate indeed, because I do not wish to be killed!

'But I'm desperate too,' said the farmer. 'I badly need to earn a decent living to support my family. You are only a bit of wood trimmed with leaves and flowers. Whereas *I* am a human being – the most important kind of living thing in the world.'

Don't overestimate your status within creation. It's true that you humans have the ability to walk all over the Earth and shape it to please yourselves, but that is nothing to be proud of. We trees, on the other hand, would never harm the Earth. For we are the living link in creation. Our branches stretch up to the heavens above, while our roots coil down to the underworld beneath us.

'So you say,' said the farmer sceptically. 'But the fact is that you're right in my way, which is a great nuisance. I need the space taken up by your roots to plant my crops. And I need to get rid of the shadows cast by your branches and leaves, so that the sun and rain can get through to help them grow.'

Leave me be, work around me, get to know me properly. That way you will learn that even wild things can prove invaluable to humans.

The farmer stood staring at the tree for a long time.

Finally, he put his axe aside and instead took up his other tools. With these, he started tilling the bare soil in the clearing – taking care to keep well away from the roots of Ciricote Tree. When this was done, he planted neat rows of seeds there, then went home.

For the next few months, he tended the seeds carefully and was rewarded by a set of strong, rapidly ripening crops.

The Ciricote Tree never spoke to him again. But when the wind blew, it dropped some of its leaves, letting them fall by his feet like a gift. He took them home to show his wife who, of course, already knew all about his strange encounter with the tree.

'These leaves look as if they could be useful,' she said thoughtfully.

Sure enough, she quickly discovered that their rough surface was perfect for scrubbing out burnt cooking pans. In fact, the sap oozing from them was just as good as the expensive cleaning fluid she'd previously had to buy from the local market.

The year turned and the tree's beautiful orange flowers developed into luscious fruits. The farmer's wife beat them to a pulp and sweetened them with wild honey to make a thick, delicious drink.

Then the farmer noticed some bark peeling away from Ciricote Tree's trunk. He took that home to his wife too. She sniffed it, squeezed it between her fingers, tasted it gingerly – then boiled it up with water to brew a tea. She gave this to their children who had all been suffering from bad coughs and stomach upsets. Within a few days, they were all feeling completely better.

Thus it was that the farmer learned an important lesson. It's all too easy to destroy wild plants just because they get in the way of human plans. But give them a chance, and it often turns out that instead of being a nuisance, they are actually invaluable.

Live in Harmony with the Wild

Bangladesh and India

Down in the mangrove swamps, a young boy lived with his widowed mother. It was a hard life, for he had to work long hours herding cattle. The two of them were very poor, with scarcely enough food to eat.

One day the boy's uncle said, 'Why don't you come and join my men collecting honey in the forest? It sells for a high price in the market, and I'll give you a fair share of the profits.'

The boy was very excited. However, his mother was strongly against the proposal, despite the promise of good money. 'I'm not having my only child roaming about in the forest!' she cried. 'It's far too dangerous. There are all kinds of savage beasts and outlaws lurking there ... and evil spirits, especially the Tiger-Demon.'

'Who's he?' the boy asked.

His mother shuddered. 'He's proclaimed himself king of the forest and regards all of us human beings as his enemies. If he catches anyone hunting – or even just collecting wild foods like your uncle's honey – he accuses them of stealing his own possessions. He's declared that he'll only allow people to do this, if they pay him a tax – a terrible tax!'

'Terrible in what way?' asked the boy.

His mother replied, 'They must let him seize whichever intruder takes his fancy – then kill and eat him.'

'But he hasn't killed Uncle,' the boy protested. 'And *he* goes into the forest regularly. This is the first chance I've ever had to earn a proper living. It's not fair to stop me taking it up.'

'Oh ... very well,' she said reluctantly. 'But promise me this, son. If you do find yourself in trouble while you're in the forest, be sure to immediately call out the name of Ma Bonbibi.'

'Ma Bonbibi?' said the boy. 'Who's she?'

'Haven't you heard of her?' she said in surprise. 'Bonbibi is the Mother of the Forest, the great protector. She's even more powerful than the Tiger-Demon. Hopefully, if you shout loud enough, she'll hear you and come to the rescue.'

The boy promised to follow her advice. Then he ran off excitedly to join his uncle and the team of honey collectors.

He had never been properly inside the forest before. It was like no other place he had ever seen: damp and beautiful, ancient and eerie. Some trees had thrust themselves up from the sodden, leaf-littered earth. Others grew actually in the water. They formed an endless muddle of tangled roots and lush greenery, stretching as far as the eye could see. They reached up to the sky and trailed heavy branches along the ground, arching over the rivers, fringed by gigantic, draping ferns. Sunlight trickled uncertainly through the leaves, scarcely penetrating the shadows. Myriad streams flowed among them, the water constantly drawing backwards, holding its breath, then rippling forwards again, hinting of floods. Monkeys scampered about, chattering loudly, and groups of spotted deer passed by. The boy was entranced as he followed the other men along a rough track, going ever deeper inside.

Every so often, they would halt and peer around at the trees. At length, his uncle said gravely, 'I wonder why none of the beehives are hanging from the trees where we usually find them.'

One of the men suggested, 'Perhaps we should turn down one of the minor paths and search there.'

Before the uncle could answer, a fearsome sound came booming through the undergrowth. It was a roar, sonorous as a heavy drum, turning into a growl of words:

'*RRRRGH!* Don't move! Stay where you are!'

Leaves rustled.

Several of the men cried out: 'Tiger!'

The mighty beast came striding towards them, muscles rippling. It was slow and ponderous, huge and powerful in the dappled light, whiskers twitching proudly. Its yellow eyes stared them all down one by one.

The men huddled together in a defensive circle, all facing outwards, pulling the boy in to join them. Birds scattered, shrieking alarm calls; the ground and water seemed to sway.

The tiger advanced on them and stopped a single stride away. It roared again, then switched effortlessly to human speech:

'Kneel before me, intruders. Pay me homage!'

At once, the men all fell to their knees, quaking with fear. The boy copied them. All around him, hoarse voices murmured:

'It must be him!'

'The Tiger-Demon!'

'How dare you enter my forest without permission!' the Tiger-Demon snarled. 'What are you trying to steal from me? Where is my payment? Which of you is the leader? Answer me!'

The uncle said nervously, 'Great lord and majesty, I assure you, we are not stealing anything. We are just humble honey collectors, coming to accept the bees' generous offer to share their surplus. Our ancestors have been doing this since time immemorial and we have never knowingly caused any offence. We did not realize you now charge a price for performing this simple task within your realm. Please tell us what we owe you, and we will pay it at once.'

'Of course there is a price,' said the Tiger-Demon. He padded round to where the boy was cowering and placed one of his huge front paws on the boy's shoulder. '*He* shall be your payment.'

The uncle blanched. He had brushed off the mother's misgivings about the dangers of the forest – and now he had led her son into the very catastrophe she had feared! His mind whirled … but on the other hand, surely it was forgivable to sacrifice a single puny boy, to save the lives of ten strong working men?

He said meekly, 'Um … of course, my lord. Will you take him now?'

'Not straightaway,' said the Tiger-Demon. 'I want some fun before I tear him to pieces. You see that boat moored by the wide stretch of water over there? Use it to row him across to the island in the middle, then abandon him. Once I've worked up my appetite, I shall swim across, chase him around for a bit, then slowly tear him to pieces.'

Reluctantly, they obeyed his command. The uncle was shamefaced but resolute. Once they reached the island, he warned the boy: 'I'm afraid there's no point in even attempting to escape, because the Tiger-Demon employs a whole army of invisible spirits. They'll be spying on you even now – and if they see you trying to break the agreement, they'll tell the Tiger-Demon – and he'll make your death a thousand times more painful than he's already threatened!'

The boy wretchedly watched his uncle row back to the shore, and lead the men on their shameful trek back to the village. He sat down on the deserted beach. There was nothing he could do but await his grim fate.

Time went by. The sun began to sink.

Suddenly, he heard splashing, moving towards him from the far shore. And there was the great orange and black head of the Tiger-Demon in the water, displaying his long fangs! Steadily he swam closer. Behind him came a murky, flickering retinue of spirits.

Death is coming, thought the boy. He braced himself. Every day of his short life seemed to flash before his eyes, sights and feelings, sounds and voices … Ah, his mother's voice! What was the last thing she said to him? *If you find yourself in trouble, immediately call out the name of Ma Bonbibi.*

Was it too late? But this was his only chance: he must try.

'Ma Bonbibi!' he shouted. 'Are you there? Come quickly if you are, I beg you – I desperately need you to save me. Please, Ma Bonbibi!'

The splashing grew louder. The Tiger-Demon was only moments away. The boy stood there, frozen with dread.

Then, soft as a breeze, he felt a strange presence embrace him. His taut muscles relaxed, and his head filled with light.

'My child,' said a woman's voice. 'I am here for you. Arise. Turn round slowly so you are not dazzled.'

He did as she said, and saw her. She was pure radiance. Her skin was pale and smooth as milk. Glossy black hair curled over her shoulders, coiling to her ankles. Her clothes were spun from golden silk. Her bangles, necklaces and anklets shimmered with the colours of sky, autumn leaves and summer flowers. Her eyes and smile were lit with kindness.

Yet her posture was that of a warrior.

She touched him lightly, then strode down to the shoreline at the same moment as the Tiger-Demon reached it.

Ma Bonbibi drew herself up and seemed to grow. Had she really stretched herself to be taller than the forest trees? Her golden robe sparkled in the dying light and the illusion faded.

The Tiger-Demon hauled himself from the water, shaking out the droplets with a roar. Behind him, his half-invisible spirit retinue loitered in the waves.

'Move aside!' he roared at Ma Bonbibi. 'Let me take my promised payment.' He licked his lips and clawed hungrily at the sand. 'Mmm! He looks a tender morsel.'

He turned his gaze to Ma Bonbibi, who stayed motionless. The boy was trembling uncontrollably. Surely at any moment, the Tiger-Demon would attack and kill them both!

But Ma Bonbibi vibrated with power, like a long drawn-out musical note. She shimmered and for a moment seemed to transform to a glistening rock, then back again, as the sound faded.

The Tiger-Demon cowered, then slowly backed away. Behind him, his spirit retinue shuddered *en masse*, stirring up the water.

At last, Ma Bonbibi spoke: 'You claim to control this forest, sir. But you well know that it is really *my* domain.' Her voice was much stronger than when she spoke to the boy, sharp as the blade of a sword. 'So I offer you a choice. If you insist on continuing as you are, threatening every human being that rightfully enters it, then beware. For I shall snap my fingers and summon my brother with his invincible army. They will come before you have time to blink, and slaughter you on the spot.

'Alternatively, you can free the boy at once, then beg my forgiveness for all your past evil deeds. Do this and I will readily spare your life.'

There was a long silence. The Tiger-Demon backed further away. He paced up and down, up and down, padding heavily over the yellow sand, shaking his splendid head from side to side. At last he returned to stand before Ma Bonbibi.

'You put this to me as if it were a simple matter,' he growled. 'You speak as if my deeds in this forest are indefensible, but that is not the case. Now, listen carefully, for I speak not just for myself, but on behalf of all the wild creatures and plants who live and grow here. Do you not know, lady, that they have willingly accepted my rule to safeguard them? Does it not occur to you, that I attack human beings who trespass here for the most noble reasons? Just supposing I were to permit them free rein here. Within a few years they would completely destroy the whole forest. The honey this boy's companions are trying to steal is only a small part of it. Their neighbours have such skill and so many weapons that they can easily kill all the birds and animals – even me. They will chop down the trees for building and firewood. They will drain the swamp, destroy the wild plants that grow there and plant it with monotonous crops. Within a few years, there will be no forest left at all. That is why I threaten and punish human beings.'

Now it was Ma Bonbibi's turn to consider things in silence.

At length she said, 'You have spoken truly, sir. So the correct way forward is to negotiate a new relationship between the wild ones and human beings, based on fairness and justice.'

She turned to the boy. 'Young friend, your companions have long gone home, so you must speak on their behalf. Step forward and stand face to face with the mighty Tiger-Demon. Do not be afraid; I will not allow him to harm you.'

The boy obeyed. With Ma Bonbibi's glow cocooning him, his fear had already melted away.

'Tiger-Demon, sir,' she said. 'I accept your role as protector of this forest, but only on the following condition. You must acknowledge me as the forest's mother, and bow to my maternal will.'

'I bow to no one,' he growled. 'Not even to you, great lady – unless human beings immediately stop their destruction.'

'That is fairly spoken,' said Ma Bonbibi. 'So, young man: you shall be released back to your village, but you too must obey a condition. You must tell your people to completely stop harming nature and the wild.'

'I solemnly promise I'll do that, lady,' he said. 'When they hear that this order comes from you, I am sure they will obey at once.'

'So, sir,' said Ma Bonbibi to the Tiger-Demon. 'You have heard him. If human beings keep their part of the bargain, will you now cease your attacks and threats against them?'

'Great mother,' he growled, 'Mighty as I am, I lack the power to refuse you. I must, so I will.'

Ma Bonbibi smiled and reached out her arms so that she was touching the Tiger-Demon with one hand, and the boy with the other. 'You both please me,' she said. 'Now then, young man, give your people this instruction too: From now on, they may only enter the forest if their hands are empty of weapons, and most of all if their hearts are pure.'

She turned to the Tiger-Demon. 'And you, sir, must henceforth welcome human visitors, since they are now rendered harmless. Your heart must be as pure as theirs.'

Reluctantly, the Tiger-Demon growled his agreement.

Ma Bonbibi showed the boy the way back to his village. Everyone there was astonished and delighted at his miraculous survival and return. They took it as a divine sign and readily believed his extraordinary tale, embracing Bonbibi's command with great enthusiasm.

After that, luck mysteriously changed for the boy and his widowed mother, lifting them out of poverty. When unexpected riches came to them, they sent up prayers of thanks to Ma Bonbibi, and were generous to all their fellow villagers.

Since then, the people of the region have always treated the forest with the care, respect and reverence that Ma Bonbibi taught them it deserves. They never take any more from it than they genuinely need.

Clear Up Your Own Mess

Ancient Babylon and modern Israel

A man made a lot of money and used it to buy a large estate on the edge of town. He built a fine house there, but what interested him most was the land around it. He planned to landscape this and turn into a beautiful garden.

Years of neglect had left this land completely overgrown with weeds, so he hired a team of men to dig them up and burn them. Unfortunately, that revealed a worse problem. For once the weeds were removed, they discovered that the ground was full of stones – making it impossible to put in any plants.

So he ordered his team to start at the far end and systematically remove the stones, one by one.

'What shall we do with them, sir?' the head gardener asked. 'They're not big enough for building a wall or a shed, and they're too rough and uneven for a gravel path. Shall we dig a hole in the corner and bury them?'

'No, no,' said the rich man. 'I don't want my new garden spoilt by ugly rubbish holes. Just throw the stones over the wall.'

The head gardener stared at him in surprise. 'But there's a public road on the other side of the wall, sir,' he protested. 'Loads of people walk down it all day long, and it's regularly used by horses and wagons. If we throw the stones there, they might hit someone. Besides, they'll make a terrible mess. It'll block whole sections of the road. There could be some nasty accidents.'

'Oh, that's nothing to do with me,' said the rich man dismissively. 'Out of sight, out of mind, that's my motto. I'm only responsible for my own land – on *this* side of the wall. If the people on the far side don't like the mess, let them clear it up.'

The head gardener shrugged. Though he felt uncomfortable about this, he certainly wasn't going to risk losing his job by arguing. So he went and passed on the rich man's order to his team.

The garden began to take splendid shape with shrubs, flower borders, water features and shady trees.

One day the rich man was surveying it, standing just inside his gate with his back to the road. A short distance away, one of his workmen was busy hurling the latest batch of excavated stones over the wall. They landed on the road with a series of satisfying thuds.

Suddenly, the rich man felt someone tap his shoulder. He spun round and saw an elderly gentleman standing on the far side of the gate, leaning heavily on a walking stick.

'How dare you touch me!' the rich man cried. 'Get away from my property!'

'I'm not inside your property and I have every right to be here,' the elderly gentleman replied. He pointed at the stones scattered all over the road. 'But *you* have no right to be so irresponsible, leaving that mess over there. I've come to make a formal complaint to you.'

'The "mess" as you call it comes from my land, so I can do whatever I want with it,' the rich man retorted.

'Really?' said the elderly gentleman. 'Have you not seen the terrible effects of your careless attitude? This road is a public right of way – which means that the entire length should be accessible to everyone. But how are people supposed to get along the section under your garden wall? Thanks to your rubble, it's almost impassable. Look at it!'

'So what?' said the rich man, steadfastly refusing to look.

'So people are injuring themselves because of you,' said the elderly gentleman. 'I myself never needed to use a stick until I slipped on your mess. As a result, I broke my leg, and the doctors say I'll never be able to walk unaided again.'

The rich man shrugged. 'As you say, it's a public place, so the authorities are responsible for clearing it up. Go and complain to them.'

'Have the courtesy to turn round and meet my eye, young fellow,' said the elderly gentleman sternly.

Reluctantly, disdainfully, the rich man did.

'I just want to say this to you,' said the elderly gentleman. 'Nothing is permanent and things you do now may come back to haunt you later. You've ruined this part of town for your own selfish gain, but one day you may regret it.'

With those words, he turned and hobbled slowly away, slithering about on the discarded stones, groaning and squealing as he searched for a safe path between them.

'Get lost, you stupid old louse, and good riddance!' the rich man yelled after him.

Then he marched up to the head gardener and said angrily, 'Why are there still so many stones left in the soil? Get rid of them, every last one. I want to see your men working even harder at throwing them over the wall.'

The garden was finished at last. It won endless admiration from the rich man's cronies.

But as time went by, the rich man's business began to falter. He owed many people large amounts of money, and the only way he could repay it was by selling off portions of his precious garden. Eventually, he did not own any of it any more.

Finally, his business failed completely. A judge ordered him to sell his house, and everything he owned as well – even his shoes – in order to settle his massive debts.

He himself was an elderly man now – and was left with nothing. From now on, he would have to live as a vagrant, by begging.

For the last time, he walked wretchedly out of his door, past the area that had once been his garden. The people who bought it from him had turned it into a building plot, and the former flower beds were now piled up with bricks. The wall was still there, but the gate had been removed. He passed through the opening, onto the road.

Until then, he had only ever travelled along this road by carriage. How he used to curse and swear whenever the horses swerved or stumbled – even though it was no fault of their own. He had always kept the curtains drawn over the carriage windows, to avoid looking at the mess he had dumped out there.

But now he had to walk along the road, barefoot, he quickly discovered what a dangerous place it was. For stones were haphazardly scattered and piled up everywhere on this section of the highway: big ones, small ones, pebbles, lumps of sharp rock and tiny shards of gravel. He had not walked very far before his feet were cut and bleeding, with pieces of grit sticking painfully into his toes. Every so often he lost his balance and toppled over, cutting his hands too, and covering himself with bruises.

Unwanted memories came crowding in on him. He recalled the callousness with which he had ordered his workmen to fling stones over the wall. He greatly regretted stubbornly ignoring the concerns of his head gardener and abusing the elderly gentleman.

For he saw now that they were absolutely right: *everyone* – including the rich and powerful – has a duty to clear up their own mess.

Leave Some for the Future

Ainu People (Japan)

Spring had turned the landscape green, and the sun was shining, so the village women went out to the mountains to gather wild plants for cooking. Most of all, everyone hoped to find some *pukusa*, for nothing is more delicious than soup flavoured with its rich, garlic-scented leaves.

Among the gatherers was the chief's new young wife. As she watched the others rapidly filling their baskets, she began to panic. For she was the most high-status woman in the community; how shameful it would be if they returned home with far more leaves than her!

She thought hard for a long moment, considering how to outdo them. Aha! That was it!

Instead of laboriously plucking the leaves one by one like the others, she took up a sharp stick and started using it to dig out entire *pukusa* plants by their bulbous roots. It was so much quicker. In no time at all, she had filled her basket right to the top. She sat on a rock and called out to the others: 'Get a move on! Look at *my* huge harvest.'

The younger women were very impressed and made to copy her method. But the older ones all scolded them: 'No! Stop! You must always leave the main part of the plants behind. If you don't, they won't sprout and grow new leaves next year.'

Of course, the chief's wife already knew that, for her own mother had taught her the correct way. She had deliberately ignored this tried-and-tested advice … but so what? There would be plenty of new plants growing next year in other nearby places. She held her head high as she led the way back to the village, displaying her overflowing basket.

When she got home, she carefully separated the leaves from the bulbs, and threw away the latter. Following custom, she laid aside most of the leaves for drying. But she chopped up the best ones to use immediately, stirring them into a piquant stew to share with her husband.

He knew nothing of her reckless behaviour, and they both heartily enjoyed their meal together.

The following day, while the chief went out to conduct his village business, his wife fell ill. First, she became nauseous and feverish. Then she was violently sick. By the time the chief returned home, she had turned as pale as bones and was shivering all over.

The chief was so concerned that he sent to another village for a healer. This was a woman renowned for her great shamanistic powers. She examined the chief's wife, then questioned her carefully.

'Have you done anything to offend the spirits?' she asked.

'I always take care to carry out the correct rituals,' was the wife's evasive answer.

'But I've heard alarming gossip going around,' the shaman pressed her. 'They say that when you were gathering *pukusa* leaves, you pulled the entire plants right from the ground. Is that true?'

Mortified, the chief's wife nodded.

The healer said, 'Then I'm afraid there's nothing I can do to cure you. The *pukusa* are very generous when they offer up the goodness of their leaves for us to enjoy. Yet instead of respecting them in return, you callously uprooted them. In doing this, you killed not just the plants themselves – but also the spirits that live within them. This is an unforgivable offence against Heaven. The illness you are suffering is your rightful punishment. I fear it may lead to *you* dying, as well as the plants.'

Don't Waste Food

Ojibwe people (Native American and Canadian First Nation)

One year the women produced the best crop of corn that anyone could remember. It grew really thick, higher than the tallest man in the village, with cobs almost impossibly fat and juicy.

The villagers ate corn soup, corn porridge, corn bread and corn cakes until they were all twice as fat as previously. Yet still there was an extraordinary amount left. They fed it to their dogs – until it made them so sick that they turned away when offered more. A huge amount was left in the fields waiting to be harvested. The women brought some of it in, but there was far too much to eat.

They ended up burying the remainder in the ground, or simply leaving it in the fields to wilt. No one could face eating any more corn for a very long time. So they didn't even bother to cut down the stalks, or weed the fields, or collect the seeds for next year's sowing.

Instead, the women now spent all their time on handcrafts, while the men went hunting for fresh meat.

Unfortunately, the men's hunting yield was far worse than the women's corn yield. In fact, to their great shame, they failed to kill anything. There were plenty of deer and just as many elk as ever, but every arrow shot went astray, and the herds all fled with twice their usual speed. So the men ingloriously returned home with no meat at all.

Hunger threatened. What an unexpected change of fortune!

The women now realized they should go to the fields to gather in the neglected remains of the last harvest. However, by that time it had all rotted away. So they hurried to the places where they had buried some of the leftover corn – only to find that mice had already discovered and devoured most of it. The few remaining cobs were totally inedible, covered with nibble marks and droppings.

The hunting continued to fail. Despite the over-abundant harvest, now they had no food of any kind at all.

The priests and priestesses consulted urgently. The only solution they could come up with was for everyone to perform the sacred rituals of singing and praying.

There was one man in the village who, from the very start, had been appalled at the over-indulgence of corn and its wastage. He had refused to take part in it.

Now, instead of joining in the official prayer rituals, he slipped away and walked off alone into the forest, thinking hard, trying to make sense of what had happened. He trekked on and on, far beyond the familiar hunting tracks, ever deeper into the trees.

After a long time, he reached a clearing. On the far side of it stood a tumbledown birch bark hut. Coming from inside the hut he heard heart-wrenching groans and cries of pain. He hurried closer, knocked, but received no answer. So he pushed open the door and peered in.

The hut was totally unfurnished apart from a few dirty, tattered hides piled up against the far wall. Lying on top of these was a stunted, wizened man, emaciated and pallid. He must surely be seriously ill. As the light from the open door fell on him, he let out another groan, then turned to examine the intruder.

'*You!*' the stunted man gasped huskily. 'You're one of *them*, aren't you? Come to complain to me, no doubt, eh? Well, it's not in my power to make things better. Only your lot can do it.'

'Forgive me, grandfather,' said the village man courteously. 'I have certainly not come with any complaints. I found my way here by pure chance. But who are you? I would dearly like to help you, so tell me what has made you ill, and what you need.'

'It's not "what" but "who" has made me ill,' the stunted man snapped back at him. 'As I've already said, it's your lot who are to blame. After all these years that I've been the best of friends to them! Haven't I always ensured that you have enough to eat? I didn't expect any thanks for it – just to be treated with respect. And this year, I've been over-generous. But what have you lot done in return? You've allowed your dogs to chew me to pieces then vomit all over me. You've thrown me away, buried me in filthy mud, let mice desecrate me. You've left me in the fields to turn stinking rotten. Look at the state of me, young man! This is all the result of your people's misdeeds. I'm left with no water, no clothes, not even a leaf to protect me from the coming cold. Soon the wild creatures will break in here and gobble me up. And once that happens, it will be impossible for you people to ever benefit from my goodness again.'

'Just tell me what to do to save you, grandfather!' the village man cried in desperation.

The stunted man had a long fit of coughing. When it passed, he let out a low moan, then at last managed to speak again in a broken whisper: 'I'm so far gone, there's only one thing you can do now. Go back to your village, tell them you have met the Corn Spirit. Then repeat everything I've just told you.'

With these words, he sank back onto his squalid bed and spoke no more.

The village man hurried back through the forest, stumbling, losing his way then finding it again, never once stopping until he was back at the village. He found everyone there still in the midst of the prayer rituals. He burst into the pulsating sacred circle, elbowing people aside, shouting at the top of his voice: 'Stop! You're wasting your time!'

The elders were outraged. They had him seized and dragged away. But he struggled against his captors shouting: 'Release me and listen! I have just met the Corn Spirit!'

At these last words, the priestesses and priests stepped forward into the centre of the circle and held up their hands. The dancers stopped moving; the singers and drummers fell silent.

The young man continued, 'The Corn Spirit is suffering unbearably! He told me that the cause of this is our irreverent wastage of the abundant gifts that he bestowed on us.'

Everyone listened quietly while he described what he had seen and heard in the forest. No one challenged him; everyone believed him. Even more important, they all understood that they must act without delay.

There were no more prayer rituals, only practical action. The women hastened to their fields, where they worked from sunrise to sunset, day after day, weeding and tilling the soil. Then they went to the muddy places where the mice had excavated the buried corn waste and managed to retrieve a fair quantity of undamaged seeds. These they carefully planted, taking great care not to spill any – for there were scarcely enough to cover the bare ground.

Meanwhile, the men went hunting again. This time they succeeded in shooting just enough game to see them through the coming months.

Winter passed. Spring came and the seeds sprouted. All through summer, the corn plants flourished. When the crop ripened, the men worked alongside the women, harvesting it carefully. They ate no more corn than they needed that year. The remainder they stored, so securely that no wild creatures could possibly damage it.

In this way, the Corn Spirit was appeased.

The people took great care never to inflict suffering on him – and thus on themselves – ever again.

Take Responsibility

Bali Aga people (Indonesia)

I'm sure you know what a nuisance wild monkeys are when they break into your house. They cause absolute havoc!

Well, the one I want to tell you about used to visit a particular village family every day. He would sneak in surreptitiously, then lurk in the shadows waiting for his chance.

The young woman who lived there used to boil yams for her elderly parents to eat. They had lost most of their teeth, so that was the only food they could manage. However, boiled yams were also the monkey's favourite, so once he started visiting, on most days they scarcely managed to get even a taste. The monkey would squat surreptitiously in a dark corner, watching while the food was being cooked. He was so quiet that no one realized he was there. But as soon as it was served, he would leap down to where the old couple were sitting, and snatch away the bowl right in front of their eyes. Then he would stick his head into it, slurp up the food in an instant, toss the bowl to the floor and scurry away, whistling and squealing happily.

Of course, they did their utmost to keep the monkey out of the house, trying one trick after another. But he was far too quick and much too cunning. They never managed to outwit him, and they could never catch hold of him, either.

So that poor old couple were constantly hungry. They grew so thin that their daughter worried they were wasting away.

In the end, she decided to confront the monkey. So one day as he snatched the food bowl and darted away with it, she yelled after him, 'Drop it, you thieving little pest! This food belongs to us. How dare you steal it!'

The monkey stopped abruptly. He put down the bowl, taking care not to spill it. He sat back on his haunches, staring round at the family with his unnervingly shrewd yellow eyes. He took his time to lick some smears of yam off his fingertips. Finally, he answered clearly, in perfect human speech:

'What a cheek! You accuse me of being a thief – when the truth is that *you* lot, you heartless people, have stolen all my usual food. Not

just mine, but also the food of all the other wild creatures that used to live in this forest.'

At first, the young woman was too astonished to answer. But eventually she stammered, 'Wh-whatever do you mean? W-we have absolutely nothing to do with any wild creatures. We never even see any – apart from you. So how can you accuse us of stealing their food?'

'What idiots you human beings are,' the monkey screeched back at her. '*Pitiless* idiots! Don't you understand? We wild ones have all gone hungry ever since your men started cutting down the trees for wood. That includes your own husband, foolish lady. Logging here, logging there, every single day. Felling trees, dragging them away, even burning the bits they don't want ... that leaves nothing for us to eat, not a thing. No flowers, no roots, no bark. Not even any wild eggs or chicks or lizards, because there's nowhere for them to shelter. And you've destroyed all the other wild ones' homes too. That's why you never see them these days. You've turned our beautiful ancestral forest into a barren wasteland – just so you selfish humans can get rich and fat!'

Now the old father spoke up. 'I used to work in the forest myself, little fellow. I assure you that it never occurred to any of us that logging could cause any harm.'

'Well, it's about time you human beings took responsibility,' the monkey retorted. 'You've created a catastrophe round here. We wild ones are completely innocent. Since there's no food left outside, we have every right to share your meals. You'll have to learn to live with it!'

Protect the Water

Trinidad and Tobago

Be careful.

If you burn down the forest trees, it's a crime. If you poach animals or kill them just for the sake of it, it's a crime. And if you foul the rivers, lagoons or mountain pools, you'll find yourself in the biggest trouble of all.

Because *she* will come for you. She'll catch you, she'll punish you.

Yes, her: Mama D'lo, Mother of the Water, great protector of the water creatures and their homes.

Oh, you foolish men, callous hunters, careless polluters! You might think you can easily evade her, but you're wrong. For Mama D'lo is the world's greatest expert in trickery.

She'll lure you with her sweet singing. As soon as you hear it, you won't be able to stop yourself: you'll follow the sound until you see her in the distance, sitting on the water's edge, tending her hair with a golden comb. In the flickering sunlight she'll be exquisitely, unbearably beautiful. You'll draw near urgently, desperate to take her …

But as you get closer, you'll suddenly hear a *CRACK*. Another and another, growing steadily louder.

It'll be her *tail*, slapping against the water … no ordinary tail, but that of a green anaconda, a giant water snake.

That's her: half-woman, half-anaconda.

Do you think you could escape her? Maybe, if you had your wits about you. But that's not very likely once you've fallen under her spell. Some people say it might be possible if you remove your left shoe, turn it upside down and walk away from her backwards …

But more likely, you'll be frozen with fear.

Then she'll whip you with her anaconda tail, wrap it tightly round you and drag you down into the depths. She'll either kill you at once or keep you in her power – the power of a snake – for the rest of your life.

That's the punishment everyone deserves if they taint her watery realm.

Keep the Song Alive

Mbuti people (Democratic Republic of the Congo)

Hush, listen. Can you hear the forest?

Leaves dropping, softly, softly; rain pattering on the leaves. Creatures scuttling, creeping. The cracking of twigs. Monkeys chattering, bees humming. Lianas creaking as they sway in the wind. Far, far off, an elephant calls.

Listen to them all, my friend. But there is one sound you will never, ever hear, because it has been destroyed forever, it is extinct. And that is the Most Beautiful Song in the Forest.

Not so long ago, this song still lived, and a young boy heard it. It was so unbelievably sweet, exquisite … how can mere words describe it? It made the boy want to cry with bitter grief and dance about with ecstatic joy, all at the same time. It was full of everything: love, longing and despair; the stab of a knife, the honeyed scent of flowers … everything, absolutely *everything* was contained in that song.

As soon as the boy heard it, he had but one desire: he must find the singer.

So the boy walked through the soft green lights of the forest, following the sweetness of the song. Ahead: it always seemed to be just ahead of him. But at last it came to rest in a clearing. And there on the ground stood the singer: a small, very plain brown bird.

The boy whispered to the bird, 'How wonderfully you sing!'

The bird was silent now. But it did not fly away. Instead, it fixed the boy with its bright eye.

The boy whispered to it again, making gentle, clucking noises like a mother to her baby. He reached out to a tree, plucked a handful of seeds from it and held them out in his hand. The bird flew down and took the food. The boy offered more. They went on like this for some time, over and over, until at last the boy and the bird were friends.

'I wish you would come home with me,' said the boy.

He spoke from his heart, and the little bird understood. With a soft fluttering of its brown wings, it perched on top of his head. And there it stayed, as the boy walked happily through the forest, back to his village.

His father was waiting for him there – and he was angry.

'Where ever have you been?' he scolded. 'Your mother thought you were lost.' And then: 'What's that stupid thing stuck on top of your head?'

The boy only answered, 'Listen!'

Almost at once, the little bird began to sing. Pure as spring water; deep as the oldest mysteries, sweetly strong like the rising sap of a thousand ancient trees. For a few moments the boy's father seemed to forget his irritation, and tears even came to his eyes.

But as soon as the bird fell silent, his face clouded again. 'Yes, very nice,' he snapped. 'But I'm supposed to prepare myself for a hunting expedition tomorrow. The other men are all ready. But *I'm* not, because your mother's been sick with worry about you being lost.' His voice rose and the last words came out with a bitter shout. He turned his back and stomped off.

The little bird stared after him for a long moment, absorbing his anger. Then, to the boy's dismay, it spread its wings and flew away.

However, the next day the boy heard its wonderful song again. The sound of it made his heart soar. Once again, he spent long hours searching for it through the trees. At last, he found the plain little bird, and quickly coaxed it back to friendship with morsels of food. Again, of its free will, the bird accompanied him back to the village.

This time the father was in a good mood, celebrating the killing of two large antelopes with the other men, taking his share as they cut up the rich meat.

The boy thought this was a good opportunity. He brought the bird to his father and urged him to listen to it singing. His father didn't shout, but shook his head at the other men, and turned away with a scornful laugh.

'Let me get on with my work,' he grumbled. 'You should be helping, my son, learning to be a proper man. Get rid of that infantile pet!'

As on the previous day, the bird was greatly alarmed by his sharp words. Almost at once, it flew away.

The following day was the same. The boy spent ages seeking out the bird that sang the Most Beautiful Song in the Forest and coaxing it to go home with him. And just as before, his father scornfully dismissed its exquisite singing, and frightened the creature away.

Nevertheless, by the fourth day, the boy and the bird had built up a really deep friendship and had learned to trust and understand each other.

'My father's nasty words can't hurt you, so don't be frightened by him,' the boy told it. 'This time, stay a little longer, I beg you. I so want my mother and my brothers and sisters to enjoy the magic of your singing.'

The little bird cocked its head and stared at him as if to say, 'I will.'

But as soon as they reached the village, they found the boy's father waiting for them, with thunder in his eyes.

'There you are, just as I thought,' he snarled. 'Off with that stupid bird again, constantly wasting your time!'

He snatched the bird from his son's head and hurled it to the ground with such force that he killed it.

The boy gave a shriek of anguish. But that was not the end of his sorrow.

For the next moment, his father gave a dreadful cry of pain, clutched his chest – and he himself dropped dead.

In this sad tale there is a lesson.

When the man killed the bird, he also killed its song, the Most Beautiful Song in the Forest. In killing the song, he wounded the forest most dreadfully – and the forest took its revenge.

So don't hurt the forest, above all don't destroy it. For its pain will always come back to you.

OUR ANIMAL RELATIONS

In the Time of the Ancient Ones

Aparai people (Brazil)

In the time of the ancient ones
all beings lived together and could talk together.
The animals were the main source of knowledge.
It was they who taught the people, songs, dances, art,
technology, hunting and healing.
There was really not much difference
between people and animals.
In fact, if there was a disagreement or conflict between them,
they often shape-shifted between one form or the other.
Sometimes they even married each other,
or took care of the others' children.

Animals Have Feelings Like Ours

San people (southern Africa)

As you know, the way that people here obtained meat in the past was by the men going out hunting. They would go after all sorts of prey: antelope, zebra, giraffe, hare, lions and so on. Each had its own particular delicious taste, and their skins and other body parts were all used to make useful items; nothing was ever wasted.

The hunter in this story wanted to treat his family to a meal of baboon meat. So he went out tracking, and eventually found a big group of these animals gathered around a waterhole. Perfect.

In the usual way, he uprooted a small shrub and held it in front of himself for camouflage. Then he lay down on his belly and wriggled closer – quietly, very quietly – to get within shooting range. Which baboon should he aim for? Ah, there was an easy target: a female squatting apart from the others. Even better, she was cradling an infant. What a piece of luck! Young ones like that always yield the tenderest meat and the softest skin.

He fitted an arrow to his bow. It sped through the air like a soft gasp of breath and struck the female straight on.

Eeeghh! She gave a piercing shriek.

Immediately, the other baboons started grunting in panic. They leaped to their feet, scattering off into the distance.

Where was the shot female? At their rear, desperately trying to keep up with them, still clutching the infant under her armpit. But the poison in the arrowhead was quickly taking effect. She tottered unsteadily … slower and slower … and stumbled to a halt by a towering rock. Near the bottom of this rock was a deep cleft. She was almost done for, but with her last grains of maternal strength, somehow managed to shove her infant to safety, right inside the cleft.

But it was not quite out of danger; for the glow of its fearful eyes gave it away.

She staggered a few more paces, then collapsed. The hunter jumped up, flung aside his camouflage and hurried forward to snatch her up.

But at the same moment, one of the males – a huge, fierce-looking one – broke off from the escaping troop and ran back.

The hunter grabbed his bow again and aimed it at the male … then stopped.

For the male had reached the dead female. He was sitting on his haunches next to her, staring fixedly at the hunter.

As soon as the hunter met his gaze, he couldn't help it: he had to drop his bow and look away.

The male baboon stayed squatting by the lifeless body, peering at it, examining it.

Seeing him distracted, the hunter seized his chance. He skirted round the pair, crept to the rock cleft, reached into it and grabbed the squirming infant. It only needed a quick knock on the head, then at least he could take one small body home to his wife for butchering and cooking.

But just as he had succeeded in easing out the infant, it gave a shrill scream, echoing its dead mother.

The male jumped up and rushed at the hunter. They stood facing each other, a few paces apart. The male grunted and barked angrily, on and on … until, in the hot sun, the sounds seemed to transmute into human words:

Leave off! Isn't it enough that you've killed the mother? Don't kill her baby too. Leave the little one for me, leave off!

The infant continued to scream. The male advanced on the hunter, teeth bared.

What should the hunter do?

This: slowly, very cautiously, he backed away.

When he had moved some distance, the male calmed down.

Carefully, with admirable gentleness, he picked up the agitated infant and tucked it under his arm. Then he bounded back to the mother. He laid the infant on top of her broken, lifeless body, crouched beside them, and began to stroke them both. Back and forth he swayed, back and forth, like a man in mourning.

The hunter was badly shaken.

Why? He hadn't done anything wrong – only what he always did, working to feed his family. But he had made a mysterious connection. He could not bring himself to kill anything else, not until after the sun had set and risen again on a new day.

So he slunk home empty handed, full of shame and remorse.

Never Hurt a Fellow Creature

Menominee people (Native American)

Three girls were walking home together through the forest, when they found their way blocked by a small porcupine. It was sitting in the middle of the path, nibbling a bunch of leaves, which it clutched carefully in its front claws.

'What a fat, ugly creature!' one of the girls scoffed, pointing. 'Look at those scruffy bits sticking out all over it, and its face is completely vacant.'

'Don't say such things,' said a second girl. 'Surely you haven't forgotten what the elders told us: we're supposed to respect all animals. It can't help what it looks like. Anyway, I think it's really sweet.'

'Only *some* animals deserve respect; not stupid, ugly ones like this,' the first one retorted. 'Why doesn't it get out of our way?' She prodded the porcupine with her foot. 'Move, you lazy thing!'

The third girl giggled. 'You need to push it harder,' she said. And she gave the porcupine such a kick that it went rolling over onto its back, legs kicking helplessly in the air. It let out a long, soft cry of distress, like a baby.

'Oh, the poor thing!' the second girl cried.

'No it's not,' exclaimed the first girl. 'Phew! What a stink it's giving off now!'

'That's because it's scared,' said the second girl. 'Please leave it alone.' She bent down and tried to help the creature upright, using her bare hands – then suddenly jumped back. 'Ouch, it's sharp!'

'What did you expect?' said the third girl. 'Everyone knows porcupines are covered in prickles – another reason why they're despicable. Come on, let's punish it for hurting you. Let's pull out all its spines. That will teach it a lesson!'

'No!' the second girl cried.

But the other two ignored her. The third girl kicked the porcupine upright again, then placed her foot firmly on its back to stop it scuttling

away. The first girl crouched beside it, pulled down the end of her sleeve to protect her hand, and started tugging at its quills. They came out easily, and soon she had made a big pile of them.

All the while, the porcupine tried fruitlessly to get free, whimpering and squealing.

'Stop it, you cruel bullies, stop it!' the second girl cried. 'The sacred powers are bound to be watching. You'll not get away with it.'

She tried to push the other two off the creature, but they just elbowed her aside, laughed in her face and carried on with their cruel game. Every so often they swapped places, so that the first girl held the creature down while the third one plucked its quills. On and on they went, paying no heed to the animal's distress, their friend's pleading or to the passing time.

Finally, they had pulled out every single quill from the porcupine's back, leaving it completely naked. It shivered violently, for the day was freezing cold. Its exposed skin was puckered and full of tiny holes where the quills had been growing. This sight made the two tormenters laugh more heartily than ever.

'It looks like a fish with legs!'

'Like a wrinkled old man!'

'Like an over-stuffed, pot-bellied worm!'

At last they released the terrified creature. It let out a final cry, then ran to the nearest tree and hastily scuttled up it, disappearing into the branches.

The first girl turned to the second one. 'There! Satisfied now, you goody-goody? We've let it go free, like you wanted.'

The second girl was in tears. 'But that doesn't undo the wicked deed you've just carried out,' she said. 'The sacred powers are bound to take revenge.'

'Powers?' the third girl mocked her. 'Huh! The only power *I'm* afraid of is my father, who'll be furious if we're not back at the village before dark. Come on, we'd better get on our way.'

So they set out again. The second girl now walked well ahead of the others, shunning their company.

In fact, darkness fell much sooner than they had anticipated, for an early winter storm was brewing. A wind flurried up, whistling through the trees, and the sky blackened. Then snow began to fall, lightly at first, but rapidly turning into a blizzard. The huge flakes quickly settled along the path, and on the undergrowth on either side of it, until it was almost impossible to see which way to go.

Still the second girl strode ahead. For a while she kept turning round to check the others were behind her. But as the snow thickened on the ground and in the air, she had to concentrate solely on where she was going.

Somehow, she managed to reach the village before she gave way to exhaustion and collapsed.

When she had recovered enough to speak of what had happened, the elders shook their heads sagely. 'You alone behaved correctly,' they said. 'It is vital never to abuse any other living creature. For they are all our kin, and one never knows how strong their guardian spirits may be.'

As for the other two girls, they never made it home. Their bodies were not found until snow-melt time in the spring.

Dogs Show How to Cooperate

Sami people (Arctic Sweden)

Managing a big herd of reindeer is extremely hard work, and the ancestors had to round them up and bring them down from the mountains entirely unaided. It's hard to imagine now just how much sweat and effort it must have taken – rushing around and chasing after the reindeer, literally yelling themselves hoarse.

Then someone suggested that perhaps they should try taming one of the more intelligent wild animals and teach it to assist them.

So they asked the Arctic fox if he was willing to try. He gave them a cunning look and said he certainly was. But he quickly turned out to be a hindrance, not a help. For his only interest in gathering the herd together was to select the plumpest calves for himself, so he could kill them and gobble them up.

'Let's ask the wolverine,' suggested one of the men. 'He's clever enough.'

So they did, and the wolverine grinned at them saying, 'I'll be delighted to help.'

At first he seemed ideal. He had a real talent for directing the herd and gathering them together, baring his teeth at any that dared step out of line. In return, the people let him take down a couple of reindeer to eat himself, so long as he only chose the oldest, frailest ones.

Unfortunately, this arrangement didn't last long. Because one day, a woman was busy sewing in her tent when the wolverine lifted the flap and swaggered in, as if he had every right to do so.

The woman jumped up, arms akimbo. 'What do you think you're doing in here?' she demanded.

The wolverine didn't bother to answer, returning her question with a snarl. No doubt, since she was alone, he assumed he could get away with anything. He slunk up to the meat store, bit it open, helped himself to a large haunch and started devouring it right in front of her.

But he'd misjudged her. For the woman seized a large stick and –

wham, wham! – she thrashed him really hard. 'That's not in the terms of our agreement – it's stealing!' she shouted, 'Drop that meat at once and get out of here!'

The wolverine did drop the meat as she'd told him – then spat and vomited all over it, making it inedible. He ran out, calling over his shoulder, 'Tell your people they'll not get any more help from me, woman. I'm off to the wilderness. From now on, I'll make a point of stealing as many of your precious reindeer as I can, and tearing the lot to pieces!'

For a long while after that bad experience, the ancestors went back to managing the herds on their own.

Then one day, when the same woman was out rounding up her herd, she heard an enthusiastic barking: *Viv, viv, viv!* She looked around and saw a dog on the far side of the reindeer, wagging his tail and running up and down alongside them. He seemed to have a natural ability to persuade them to move in the direction that the woman wanted. In no time at all, the dog had manoeuvred the entire group into exactly the right place.

The woman ran up, reached out and started patting him. 'Well done!' she exclaimed. 'You've been a fantastic help, without even being asked!'

The dog turned his head happily this way and that, to show how much he enjoyed both the praise and the affection.

'Just wait there,' said the woman. 'I'll bring you a reward.'

She hurried into her tent, opened the food store, cut off a juicy chunk of meat, took it outside and offered it to the dog.

But to her great surprise, he put his ears and tail down and backed away diffidently. 'Oh no, that's *your* food,' he said. 'I've just been having fun. I couldn't possibly accept anything as fine as that.'

The woman thought for a moment, then fetched a large bone trailing shreds of meat, the sort she used for cooking soup. She held it out to the dog, and this time, he wagged his tail enthusiastically, took the bone carefully in his mouth and carried it away to gnaw. When he had eaten his fill, he sighed contentedly, lay down close to the woman's tent and went to sleep.

The next morning, he was still there. 'Anything else I can do for you?' he asked hopefully, when the woman and her husband emerged.

'There is indeed,' said the woman. 'You're such a good worker, we'd like to take you permanently into service, helping us herd the reindeer.'

The dog wagged his tail excitedly.

'The only problem is,' she went on, 'we're worried that someone with your skills and talent will want to be paid more than we can afford.'

'Oh, don't worry about that,' said the dog. 'I love both the work and your company. It would give me great pleasure to help you every day. In return, all you need give me is regular meals of meaty bones, like you fed me yesterday. Also, if you could sometimes spare a drop of soup with lots of fat in it, I'll be completely satisfied.'

And so it was agreed, there and then.

Ever since that time, dogs have been an invaluable part of Sami reindeer herding teams. They love their work. Neither side ever breaks their contract, each always fulfilling exactly what they've promised to do.

People and Elephants Helping Each Other

Ancient India and Southeast Asia

A great bull elephant was pushing his way through the forest when he trod on a large thorn. It was so long and so sharp that it pierced right through the thick skin of his foot, causing him immense pain. In the way of most animals, he stoically attempted to continue his journey. However, with every step, the thorn dug in deeper and deeper, and the pain grew more excruciating. Gradually, his foot swelled up and the wound began to fester, making him feel light-headed.

Just then, he heard voices close by, and the *thwack! thwack!* of axes. He thought to himself, 'That sounds like humans. I've heard they can be quite savage; but even so, they're my only chance of getting help.'

He dragged himself towards the noises until he saw, through the trees, a clearing where some men were busily working. 'They look much too small to harm me,' he thought, 'so long as I don't frighten them by going too close too quickly.'

The people he had happened upon were a band of carpenters. They spent their days chopping down trees, sawing them into beams and planks, hauling them down a long slope to the river, then launching them into the water to float downstream to a timber merchant in the distant town. It was gruelling work that only earned them a pittance – barely enough to keep themselves and their families alive.

As the wounded elephant approached, ripping down branches and trampling undergrowth, they spun round in alarm. They stared as the elephant staggered out before them, dragging one of his back legs, which made his gigantic frame appear grotesquely lopsided. He stopped a short distance away, making a soft blowing sound, followed by a series of squeals.

The men all threw down their axes and hastily stepped back.

'What's the matter, animal?' the nearest one called out nervously.

Slowly, the elephant turned around and held up one of his back feet, displaying the thorn.

'That looks like agony,' another man exclaimed.

The elephant lumbered down to his knees then rolled over onto his side, with the afflicted foot sticking out. The carpenters huddled together, discussing what to do. Through his haze of pain, the elephant heard them muttering:

'He seems a harmless enough beast.'

'No, the pain could make him go crazy and attack us. Look at the size of him: we wouldn't stand a chance.'

'But he seems to be trusting us, asking us for help. It would be cruel to ignore him.'

'Yes, the right thing to do is risk trusting him back.'

The elephant was now sinking in and out of a stupor. Through it, he was dimly aware of one of the men squatting beside him, brandishing a tool. He felt it stabbing into his infected foot, then digging about in the flesh. This made the pain so intense that his instinct was to kick the man away; but he controlled himself. Suddenly the thorn came free and the pain immediately lessened.

More men came, carrying bowls of clean water from the river. He felt gentle hands washing out the wound and pressing herbal leaves onto it. Moment by moment, the soreness melted away.

The elephant let out a long sigh of relief. 'What good creatures these humans are,' he thought. 'They have undoubtedly saved my life! In return, I hereby vow to make myself useful to them for the rest of my days.' Then, exhausted from his trauma, he succumbed to sleep.

The carpenters finished their day's work, putting in extra effort to make up for their lost time, for they could not afford to be underpaid. Then they went home.

The following morning, when the carpenters returned to the forest, they were surprised to find the elephant still there. Their first aid had clearly been effective, because he was now standing with his weight on all four feet, looking completely unconcerned. He greeted them with a cheerful trumpeting. They petted him a bit, then got down to work.

The elephant stood watching them intently.

Soon one of the men had chopped down and sawn up the first tree of the day. He bent over to bind the planks in a thick rope, ready to drag them down to the river. However, the elephant came up behind and nudged the man with his trunk over and over, until he moved aside. Then he coiled his trunk round the bundle, lifted it up easily and carried it down to the riverbank himself.

The woodchopper stared at him in amazement. The elephant carefully laid down his burden, waved his trunk and opened his mouth in an expression that was surely a smile. Then he walked sedately up the slope to another man. Seeing him approach a pile of tools, the elephant got there first and delicately picked up one tool after another in his trunk, holding each one out until he had found exactly what the man wanted.

Thus it went on all day. The elephant observed carefully, learned quickly, fetched things whenever they were needed and insisted on carrying out all the heaviest work. By sunset, the carpenters had fully accepted their new friend as an essential member of the team. Whenever they sat down to eat, they all made sure to give half their portion of humble food to the elephant. He accepted this graciously, supplementing their offerings with his usual diet of grass, leaves, bark, roots and saplings.

Every day after that, the elephant returned to help them. It made a tremendous difference to their productivity. For now, they could work for much shorter periods but achieve far more. This greatly improved both their earnings and their lives in every way.

As the years went by, the elephant continued to work alongside them. In return, they shared their food, made a great fuss of him, and always ensured he had plenty of free time to forage in the forest.

One morning, they were surprised to find another, smaller male elephant at his side. This was no ordinary youngster, but a rare and very handsome albino. He had pale pink skin, fair eyelashes and milk-coloured toenails.

'This is my son,' said their friend. 'Though I would happily continue helping you good people forever, unfortunately I am starting to grow stiff and slow with age. So it is time for me to retire among the trees. Before I do so, please accept this gift of my most precious offspring, who has come to take over my work.'

This announcement was greeted with cheers of thanks and exclamations of sadness for his well-earned retirement. After that, the old one started showing his son everything he must do to help the carpenters. Soon the pair were working side by side.

Gradually, day by day, the young white elephant did more, and the old one did less.

One day, the carpenters arrived at work to find that the old bull had vanished. They never saw him again.

The young white elephant worked just as eagerly and diligently as his father. Like him, he was rewarded by the carpenters' deep friendship and much more besides.

Sometimes the carpenters' children came to watch them at work in the forest. They had been very fond of the old elephant, and when they discovered that his son had taken over, they were even more excited. For the white elephant was not much older than them and shared their love of fun and mischief. When the day's tasks were done, he eagerly joined them in the river. There they playfully splashed water all over him, while the young elephant squirted them back with gushing showers from his trunk. Soon they were all frolicking around, filling the forest with laughter.

Just at that moment, the king and his entourage happened to come along the road that ran above the logging site. Hearing the happy commotion, the king ordered his men to turn aside and find out what was going on. He was enchanted by the sight of a pure white elephant playing with the children. He rode into the forest camp and asked the carpenters where they had found him. Nervously they told the great man how they had come to be blessed by elephant helpers. The king was both astonished and impressed.

'I want to buy this extraordinary animal from you,' he said.

'He's not for sale, your majesty,' the carpenters replied as one.

'How dare you oppose me!' the king cried in a fury. Without further discussion, he signalled to his servants to throw ropes round the elephant's neck and lead him away.

The young elephant knew well the story of how the carpenters had saved his father's life, and his obligation never to let them down in return. So he refused to budge. The servants tugged him with all their might, but to no avail.

Then the young elephant said to the king, 'These people are like family to me and depend on my strength in their daily work. I will never abandon them unless you first pay them full and adequate compensation. By this I mean that you must give them every single item of treasure that you have with you today.'

This fluent speech from the mouth of a remarkable animal who was clearly his own master made the king covet the elephant even more desperately. So he ordered his servants to empty all the bags they were transporting, and lay piles of gold and jewellery around the elephant's feet. It was far more wealth than the carpenters had ever dreamed existed.

The young elephant examined the hoard carefully. He thought, 'This seems more than enough for them all to permanently lay down their tools, and also raise their families' status. If my father were here, I believe he would be satisfied.'

So, amid many sad farewells, he finally trumpeted his agreement to go with the king.

Thus everything turned out very well – not just for the carpenters, who could now live in comfort and ease, but also for the white elephant.

For his new home was in the lush gardens of the royal palace, where the king treated him with infinite kindness, housing him in a splendid stable and sprinkling him daily with sacred water. Moreover, the king was so impressed by the white elephant's intelligence and noble demeanour, that he solemnly promised to leave him an inheritance of no less than half his splendid kingdom.

Learning to Live with Mouse Neighbours

Pueblo people (Native American)

At first, when a band of mice moved into the pueblo, the people just shrugged – and their children were enchanted.

'Such small, cute creatures!'

'Look at their bright eyes and twitching whiskers!'

Unfortunately, it didn't take long to realize that this tolerance was a naive mistake. For the mouse families were much larger and grew much faster than human ones. In no time at all there were hundreds – no, *thousands* – of them, overrunning the whole town and the farms that belonged to it. And their habits were not only annoying, but extremely unwholesome.

In the houses, they ran everywhere, scurrying under people's feet and littering the well-swept floors with droppings. They never cleaned up after themselves or took care of their surroundings. They invaded the food stores, nibbling some of this and bits of that, spilling and contaminating everything. Their rustling and squeaking kept everyone awake at night. They even had the insolence to scamper over beds when people were actually asleep in them. In the fields they caused untold damage to the crops, and gnawed bits out of every single piece of fruit just as it was ripening.

It was unbearable.

The women took charge. They summoned the pueblo warriors and asked them to destroy the mice once and for all. These warriors were famed for being both clever and invincible, more than well-equipped to carry out such a simple task. They organized themselves as if fighting an actual war. They strutted around, loudly proclaiming their promise to restore complete order to the beleaguered town. They stationed themselves in key areas both indoors and outside, keeping vigil with bows and arrows. Whenever a mouse dared to creep in front of them, they immediately shot it dead.

Soon they had amassed several enormous piles of tiny corpses.

They assured the people that any mice not yet killed had fled. Peace and cleanliness were restored.

What a relief! The women made the houses spick and span again. The farmers repaired and replanted their fields.

The victorious warriors carried their weapons into the *kiva,* the sacred underground chamber, to prepare a big celebration festival. Custom required the warriors to go without food for several days before it began. This ritual fasting quickly drained their energy, especially after their long, wakeful nights fighting the mice. So it was hardly surprising that they all soon lay down and fell asleep.

Now, in truth, the warriors had not managed to eliminate *all* the mice. A few still remained right there in the pueblo, lying low in concealed nooks and crannies, spying on their human enemies. It didn't take them long to spot the warriors slumbering obliviously in the *kiva*. What a pathetic sight – and what an easy target!

The mouse spies dashed out to the wilderness where hundreds of their fellow survivors were waiting and preparing to fight back. As soon as they heard the spies' news, these mice seized the tiny spears they had been making and ran back to the pueblo. There the spies led them down the ladder to the *kiva*. They were overjoyed to see that the warriors who had slaughtered their relations were now fast asleep on the floor, bows and arrows carelessly flung beside them, completely vulnerable.

Silently, moving quickly, the army of mice set to work. Their spears easily cut the feathers off all the warriors' arrows. Their sharp teeth speedily gnawed through the warriors' bowstrings.

When they had finished, in unison they squeaked a deafening *WHOOP!*

At once, all the warriors sprang awake – to be greeted by the sight of teeming mice, brandishing miniature spears! Within moments, the mice were stabbing them all over, drawing blood.

The warriors snatched up their bows and arrows – but they were now irreparably damaged. Whatever could they do? Assailed on every side by mouse weapons and teeth, the only option was to flee outside – chased by the mouse army.

When the other townsfolk saw this, they could scarcely believe their eyes. Here were their supposedly intrepid warriors, ridiculously fleeing a squealing mob of creatures smaller than their own hands! Everyone burst out laughing, calling their families and friends to hurry from the houses and fields to see the fun.

The warriors were mortified.

But the mice were elated. They gathered in the central square, lit a bonfire, donned their festive clothes and organized a victory dance of their own. What a fine sight they were, dressed in red blankets and brightly coloured moccasins, tiny leggings ornamented with silver! How exultantly they squeaked! The celebration went on for four whole days, until the mice finally marched away to reoccupy their favourite holes around the pueblo.

In this way, mice proved themselves to be more than equal to us humans. We have to accept that they're always going to share our homes.

Loyalty to a Whale

Māori people (New Zealand)

For a long time, a certain chief's greatest friend was a whale. Their affection was mutual, and the whale took much pleasure in supporting his human comrade.

This is how it worked. Whenever the chief needed meat for a feast, the whale would catch wind of this, then swim to the shore and lie quietly at the bottom of the beach. He would allow the chief to climb onto his back with a sharp knife, and very carefully slice off a large piece of his flesh to cook. Afterwards, the whale would slide back into the sea, where the clean salt water immediately healed his wound, making him whole again. On other occasions, if the chief wished to visit a different island, he would summon the whale, who was always happy to carry him wherever he wished to go.

At the beginning of their friendship, the whale had specified the correct rules of behaviour, which the chief always took care to follow. He was very protective towards his friend, and forbade anyone to harm the whale or his kin. This ensured that both greatly benefited from their cooperation.

Their friendship had lasted for some years when the chief took a wife, who soon gave birth to a son. Now the chief loved his son as much as he loved the whale, and wanted to ensure that the boy would have a successful life. So he summoned an eminent magician from across the ocean, to conduct the necessary rituals to achieve this.

After the ceremony, the chief invited the magician to join the celebration feast. The magician accepted eagerly, for he had heard intriguing gossip about the whale-meat served on such occasions, and wished to taste it himself. He was greatly impressed to see the whale come at once when the chief called across the waves to him. He was even more impressed when the proud animal lay placid and still on the beach while the chief cut a chunk from his body, then swam away completely unperturbed. The whale meat was roasted with fragrant leaves in a great earth oven. When it was served, the magician declared it to be the most delicious food he had ever tasted in his life. He was full of admiration and praise for the chief's relationship with the wild creature, and expressed the hope that this would long continue.

However, what the magician said, and what he really thought, were two very different things. Despite his fine words, he coveted the meat from this whale for himself.

So at nightfall, when the chief called for a canoe to transport the magician home, he declined the offer, saying, 'I want your pet whale to carry me over the sea.'

'He is not a pet but an equal,' said the chief. 'And he never carries anyone but me.'

'Don't insult me,' the magician retorted. 'Have I not conducted the extremely complex and mysterious rituals for your son exactly as you requested – and even more? Yes? So I deserve a special reward for ensuring his glorious future.'

Thus he went on, accusing the chief of undervaluing his talent and skills. 'It is only a trivial matter for you,' the magician insisted. 'But it is vital for me, since I must regularly experience new marvels in order to learn from them. That is how I developed my abilities to help people such as your own family; it is a never-ending process. By refusing, you not only diminish yourself, chief, but you deny others my best services for the future. I am extremely offended.' He sighed deeply and shook his head. 'If you refuse, when your next child is born, you will have to find a lesser man to bless him. I never return to anyone who snubs me.'

He continued his argument over and over. Finally, the chief was so beaten down that he submitted. With great reluctance, he summoned the whale, and helped the magician onto the proud animal's back.

'Now please listen carefully,' he said. 'It is vital to follow the rules that my whale friend himself has laid down. The only place on Earth that he can land is here, on this beach immediately below my house. As you approach your own island, you will feel the whale shake himself. That is a sign that you have reached such shallow water that the seabed is touching his belly. This means he can swim no further. At that point, your voyage will be over. You must immediately jump off his back and wade ashore.'

'Thank you for telling me this,' the magician said coldly.

With great reluctance, the chief signalled to the whale, who innocently carried away the magician on his back.

The chief watched them depart with great misgivings. Then he entered his house to spend time with his wife and son.

The chief's greatest fear was that the magician would put a spell on his beloved whale to keep him in his power. However, what happened was far worse.

For when they came to the magician's island and the whale shook himself, the magician laughed callously. As he dismounted to wade through the last stretch of water to his home, he used his uncanny powers to push the whale forcefully down, right onto the very seabed. The full weight of his magic completely crushed and trapped the whale. He struggled with all his might but could not escape. Unable to surface, he gasped for breath. His desperate thrashing about stirred up clouds of sand. This floated into his blowhole, suffocating him to death.

Of course, the chief knew nothing of this. But his sense of foreboding grew ever stronger and more worrying.

Eventually, the smell of roasting whale flesh came drifting across the ocean from the magician's distant island. Then the chief knew for sure that the worst possible thing had happened. The most noble and generous of all wild creatures – his greatest, irreplaceable friend – had died to feed the stomach and deceit of a man who was now his most hated enemy.

If a person is murdered, the killer must be punished. To the chief, the whale he had lost was more than equal to a person, so his killer must be punished too.

He sent a band of 40 fearless, cunning women to capture the magician by subtle means. They paddled a large canoe to his island, then used brazen enchantments of their own to trick their way into his house. There, they bewitched him into an unnatural sleep, and abducted him.

As soon as they brought him back to the chief, the evil magician met a fate that exactly matched the one he himself had inflicted on the whale.

Wolves as Protectors

Japan

The woman at the centre of this story – forgive me, I do not know her name – was heavily pregnant. Since this was her first child, naturally she was nervous about the impending birth.

It was the custom at that time for an expectant mother to move back to live with her parents, so that her own mother could oversee the confinement. Unfortunately, this woman's parents lived a long distance away, and the path between their homes led over a high mountain covered in dense forest. In those days the only means of travel for humble people was walking. The woman had intended to make the long and difficult journey while there was still plenty of time in hand. However, she needed her husband to escort her, and he worked for a very demanding employer who constantly forced him to delay it.

Eventually, on a bright autumn morning, the husband wrapped their possessions in a large cloth, balanced them on top of his head and they set out together. The woman leaned on a stick to steady herself in her fragile condition. They made good speed, aiming to reach her parents' house before sunset.

Halfway along their trek, the husband suggested a brief rest to take refreshments. They sat on a rock, from where they admired the beautiful view across the surrounding peaks and valleys. When they had finished their light meal, the man rose and reached out to help his wife to her feet – only to see that the colour had drained from her face, leaving it as pale as bones. She declined his assistance, put a trembling hand over her swollen belly and let out a long groan.

'Whatever is the matter, my dear?' the husband asked nervously.

The woman gasped, groaned again and said hoarsely. 'I'm feeling terrible pains, one wave after another. Our baby won't wait until we reach my parents – it's determined to be born at any moment!' She began to pant. 'I'll have to give birth *here* – and you will have to help me.'

The husband had no idea what to do and was greatly alarmed. Nevertheless, he dealt with the crisis admirably, doing everything he could to come to the aid of his wife. As the day wore on, despite the inauspicious conditions, she gave birth to a healthy baby boy. The husband found a mountain spring nearby, fetched water from it for drinking and bathing, then swaddled the child in a shawl from their luggage.

At first, they were both overwhelmed with relief and joy. But as the new mother sat nursing their son, she burst into tears. 'I'm so sorry,' she said, 'but I feel too weak to finish our journey today.'

'That's only to be expected,' he replied gallantly. 'We'll stay here for as long as you need.'

He unwrapped the cloth enclosing their luggage and spread it over the ground. She lay down on it, nestling the baby beside her, and the husband covered them with his padded jacket for warmth.

By this time, it was late afternoon, and the sun had long ago vanished behind the western peaks. The woman began to doze.

Night fell.

Suddenly, something made her jerk awake. The baby was still sound asleep, breathing softly. But by the light of a half-moon, she saw her husband standing a few paces away on the path, looking round in great agitation, brandishing a fallen branch. He had clearly spotted some kind of danger.

The woman's first terrified thought was of robbers or evil spirits. However, when she followed his gaze, she saw numerous pairs of apparently disembodied, yellow eyes. They were dotted along the path in both directions, and hovering on the forest edge.

A pack of wolves!

Since the woman was indisposed, and had her newborn to take care of, she could do nothing to help her husband repel them. Her only possible weapon was noise. So she drew a deep breath and screamed, 'Get away, be off!'

The baby stirred in her arms, then woke up and started screaming alongside her. The wolves did not move. They stood silently, staring at the baby, the woman and her husband.

Again, the woman called out, 'If you must kill us, then at least do it quickly!'

Very softly, from all directions, the wolves padded closer.

The husband roared at them and slashed his branch through the air. The wolves dodged it, paying him no heed, continuing to approach. They stopped a short distance from the woman and her baby. One by

one they sat down in a circle around the cowering family, watching them intently.

In the moonlight, the woman saw to her amazement that the animals' expressions were not at all hostile – but benevolent.

The wolves stayed there, guarding the family all through the night. Every so often, one stood up to prowl around with pricked ears, as if vigilant for unseen dangers. Then it returned to its place in the circle, eyes fixed on the new family.

When morning came, the wolves remained where they were. The woman managed to stagger to her feet and tentatively practised walking. They watched her until she nodded to her husband, who strapped the newborn to her back, wrapped up their luggage and hoisted it onto his head. The wolves moved aside to let them pass, then melted away into the shadows.

Thus, it was that the family survived their lonely nocturnal ordeal unharmed, and travelled on safely to the new grandparents' house.

When they described what had happened, their relations expressed little surprise, and assailed them with tales of their own. They spoke – sometimes dramatically, sometimes amusingly – of numerous adults and children who were said to have been saved in the wilderness by benevolent wolves.

Finally, they exhausted all the stories they knew. Then the new grandfather hushed them and shared a piece of wisdom he had heard long ago when he himself was a child:

'If you venture into the mountains after sunset, and a wolf suddenly emerges from the trees or a clump of reeds, avoid its eyes, don't panic and definitely don't run away. Just keep walking without looking back, and take care not to fall or even stumble. The important thing is to act calmly. If you manage this, the wolf will stay calm too and protect you all the way to your destination. As soon as you arrive, it will vanish.'

Learning from Bears

Wyandot people (Native American and Canadian First Nation)

A woman and her husband decided to visit some family members who lived in a different village on the far side of a lake. Their route took them up a high mountain, then down through a thick, dark pine forest. When they had been walking through this forest for some time, they suddenly heard a loud noise, as if something was falling from the slopes above. The next moment, they were surrounded by bears!

They had never seen so many bears gathered together at once, for usually they are solitary animals. As the couple gazed at them in awe, a huge male – clearly their chief – reared up on his hind legs.

'Greetings, humans,' he rasped at them. 'I order you to accompany us.'

The man quickly overcame his surprise at the bear using human speech. He stood his ground and asked defiantly, 'Why should we? Who are you, to order us about?

'You must come and live with us for a while,' the bear chief replied. 'You will learn why in due course, but I assure you it will be of great benefit to your people.'

'Leave us alone,' the man said. 'Please get out of our way.'

'No,' said the bear chief. 'You *must* come and stay with us until we have fulfilled our sacred task. You will not be allowed to leave until I give permission.' He lifted up a huge paw and beckoned the couple to follow him.

'But at least tell us what this is about,' the man insisted.

The chief ignored him. Meanwhile, the other bears had all shuffled much closer around the human couple.

'We'd better do what they say,' the woman whispered frantically. 'Otherwise, they'll probably tear us to pieces!'

Her husband reluctantly agreed.

The bear chief led the way along a long, narrow path, deep under the trees. As they got further along it, the bears' behaviour began to change for the better. They put aside their threatening manner and started chattering very animatedly in the human tongue, with cheerful bursts of laughter. They frolicked through the trees, swung from overhanging branches, turned somersaults in mossy clearings and started play-fighting each other.

The woman and her husband could scarcely believe their eyes.

At first, the bears took care not to get too close as they played around. But after a while, a cub crept up to the man and cuffed him playfully. After some hesitation, he very gently tapped the cub back on its arm. His wife froze, dreading what they would do to him. However, the bears all let out whoops of delight, and one seized the man's hand to draw him into the game. Soon he was joining in with all their rough and tumble, having more fun than he had known since he was a young boy. At one point, he was knocked to the ground; but he just grinned and leaped straight up again, earning a round of applause from the bears.

With such entertainment, the long, undulating journey seemed to pass in no time. Shortly before nightfall, they emerged from the pine forest into a valley. A range of mountains loomed ahead. In the dying rays of sunset, they appeared to be an unusual ruddy colour, like flames. The older bears all gave grunts of happiness, while the cubs jumped up and down, crying, 'Almost there!'

The woman asked where they were.

'We're about to enter the Red Mountains,' the bear chief replied. 'They are sacred to us, for they were dyed with the blood of our ancestors. No doubt you're feeling tired. Well, it won't be long now until we arrive. Then I will show you the dwelling we have prepared for you two to live in.'

With renewed energy, the bears helped the couple to climb a steep path into the Red Mountains. At length, they emerged onto a broad plateau, surrounded by steep rocks of similar colour, full of dark openings.

The bear chief took the woman's hand in one great paw, and the man's hand in the other, led them to one of the openings and took them inside. They found themselves in a deep, high-ceilinged cave.

'Isn't it splendid?' said the chief. 'You see, we have made beds and mats in the human style, with covers fashioned from our own softest fur. There's a fire already laid for you by the entrance. And just outside you will see two groves of trees, laden with the tastiest nuts in the world. They are for you to feast on to your hearts' content. I do hope you are pleased with it, for you have no choice but to live here for a while.'

'For how long?' the man asked grudgingly.

'Until we have fulfilled our mission in bringing you here.'

'Er … it's very nice,' said the woman. 'Is it …?'

'Mission?' the man interrupted her. 'What's all this about?'

'You will discover this and much more when the time comes,' the bear chief replied. 'For now, I cannot answer any questions.'

Since the only choice was to obey the bears or risk being attacked by them, the man held his silence.

Over the next few days and nights, the couple tried to settle into their new home. It was comfortable enough, within its limitations. But although they could come and go from the cave as they pleased, they were confined to the plateau in the Red Mountains. The bears treated them with both courtesy and kindness, but they felt like prisoners and longed to go home.

After some days, the man whispered to his wife, 'I've been thinking. We must try to escape.'

The woman was doubtful. But the husband explained his plan at great length, until he managed to convince her.

The very next day, it happened that the bears had all wandered right away from the plateau, foraging for river fish and berries. So the couple seized their chance. They slipped out of the cave and started down the mountain slope.

However, some of the bears were quite nearby and on alert. They spotted the fleeing couple and rushed after them on their four sturdy legs. They soon caught them up and restrained them until their chief arrived.

The chief's face was severe and his growl like thunder: 'We have given you a comfortable cave to live in. We have treated you to plentiful good food. Yet in return, you have abandoned us. I had planned to make your lessons easy, but now you must learn the hard way.'

'I'm so sorry …' the woman began.

The bear chief paid her no attention. Instead, he seized the man – and hurled him straight down the precipitous mountain face.

He pitched headlong, falling and falling past the endless red rocks. At length he came to rest on a ledge, groaning piteously, covered in gashes and bruises, his limbs unnaturally splayed.

The woman gasped and began to sob.

'Be quiet,' the bear chief scolded. 'You have work to do. Then you will understand why we brought you here.'

He called to his companions in bear-speech. At once two of them clambered nimbly down the rocks to where the man had landed. The woman peered over to watch, and felt some relief when they carefully carried him back up.

They all returned to the couple's cave. There the bears gently laid the injured man on the bed. The woman rushed over to him.

However, the chief tapped her on the shoulder, saying, 'Leave him for now and come with me. It is time for your first lesson.'

He led her outside, across the plateau and into a wooded area. There he handed her a basket and pointed to various plants, telling her to pluck leaves from some, shoots and flowers from others. He also ordered her to dig up particular roots, and scrape bark from certain trunks. All these items she must place in the basket. When it was full, they returned to the cave entrance and kindled the fire. The chief explained how to crush the items she had gathered, mix them in specific quantities and combinations with water, and brew them in pots simmering over a low flame. When these tasks were completed, he took her across to her husband. Then he instructed her on how to administer the medicines and ointments they had brewed to the man's damaged body.

As soon as this was finished, the man's broken limbs and gashes miraculously healed, and his bruises faded away. The colour returned to his face. He sat up, blinked in bewilderment and asked, 'What happened?'

'You nearly died,' said the bear chief, 'but I have taught your wife the secrets of healing.'

To the woman he said, 'Remember everything you learned today, for the sake of all your people. But I have not finished with you yet.'

After this adventure, the couple tried to settle down in the cave, and for a while the man accepted their situation. But soon his bitter resentment returned, and he urged his wife to join him in another attempted escape.

'No,' she said, 'not after what happened last time.'

'At least if they hurt me again, you know how to cure me,' he said blithely. 'Well then, I'll go alone.'

Shortly afterwards, he set off. But he hardly went any distance before the bear chief suddenly appeared, blocking his path. An angry roar quickly summoned the other bears: 'Punish him!'

At once they set about biting and scratching the man, until he was writhing on the ground, begging for mercy.

The chief gave a yelp and held up his paw, saying, 'Enough! I would rather have found a gentler way to finish their lessons, but so be it. Sometimes esoteric knowledge cannot be acquired without self-sacrifice and pain. Carry him back to his wife. I doubt she will be surprised by his humiliating return.'

Back in the couple's cave, the bears again laid the man on the bed. He was bleeding heavily, retching and groaning with fever. The chief left him to suffer while he took the woman to gather a new batch of herbs – different ones from previously. Back in the cave, he showed her how to brew these into new tinctures and remedies, and told her the correct ways to apply them.

By the next morning, all the man's wounds had again fully healed, his fever had subsided and his appetite returned. He was greatly humbled by the double punishment and made no further mention to his wife of wishing to flee.

Later that day, the bear chief paid them a visit. This time, he lingered at the cave entrance and said in a kindly tone: 'May I come in?'

The man nodded diffidently and made space for the chief to lower his great, furry bulk onto the bed beside him. The woman went outside and hurriedly returned with an offering of nuts.

The chief nibbled them thoughtfully then said, 'Listen, my friends. It is time to explain why I brought you here and why you have suffered. First, I must emphasize that we bears consider ourselves to be great friends of your people, almost like close kin. We wish to help you thrive in the future.' He gestured to the man. 'That is why we abducted you and allowed you to suffer injuries – to enable your wife to learn the art of healing.'

He turned to the woman and placed his great paw lightly on her knee. 'Do you remember which herbs, roots, flowers and bark I told you to pick?'

'Yes,' she said. 'They are all firmly impressed on my memory. Last night I lay awake for a long time, going over their names, recalling how to prepare them and what each mixture was for.'

'I am very proud of you,' said the bear chief. 'Take your new knowledge home to your people and share it as widely as you can. Teach it to your elders, to younger people and also to children. In this way, healing and medicine will become a part of human culture, passing down through the generations. It will enable many people to avoid unnecessary suffering, so they do not die before their time.'

'I will do my best,' she promised.

The bear chief turned back to the man. 'Now, I ask your forgiveness for the pain inflicted on you as part of this great lesson to humanity.'

'Of course,' the man answered respectfully.

'And you are desperate to go home, are you not?'

'Yes – but only if the time is right and you will allow us to.'

'It is and I do,' said the bear chief. 'But I have one last thing to teach you. Although we bears are almost like relations to you humans, you may still sometimes kill us for food. However, this is only when there is no alternative, and you are desperately hungry. In such circumstances, you must carry out rituals to honour our bones. Please pass on this on to your fellows.'

The man and the woman both nodded solemnly.

Now the chief's demeanour suddenly changed. He clapped his paws loudly, let out a jovial roar and cried, 'Time to be on your way!'

The other bears ran up to the cave entrance. He spoke to them in their own language, and they responded by cheering and dancing on the spot.

The chief told the couple, 'Gather up your belongings and come.'

The bears took them back across the plateau, down the rocky slopes, past the lake and through the dark pine forest. Just as on the outward journey, there was much fun and frolicking as they went. The woman and her husband could not resist joining in. As they did, the traumas they had experienced gradually faded, like the memories of strange dreams.

But in her mind, the woman clung tightly to the wonderful new knowledge that the bear chief had taught her. What an honour! She determined never to let it go.

When they emerged from the forest, the chief halted and the other bears gathered round. 'Now it is time to say goodbye,' he said. 'Thank you for your patience, my friends. I wish you and all your people well.'

The couple descended the last slopes back to their village on their own. How amazed and happy everyone was at their safe return, after such a long, ominous absence! The couple told their extraordinary story around the fires that very night, to many exclamations of astonishment.

From that time on, whenever someone fell ill or was injured, they consulted the woman. Her cures never failed. She gladly taught them to anyone else who expressed interest. In this way, many other people obtained healing skills, making the village much healthier.

Thus it was thanks to the bears that knowledge of medicine first came into the world.

Sharing a Jaguar's Power

Manchineri people (Brazil)

There was a young woman whose parents and grandparents all died before she was old enough to be initiated into adulthood. She stayed living with her two older brothers, who were both married and shared the ancestral family house. Unfortunately, neither brother had any affection for her, and both wives treated her with contempt. She was very unhappy.

One day she went down to the river to bathe. As she splashed in the shallows, she suddenly had an uncanny feeling that someone was watching her. She spun round. There was no one on the bank. However, behind her, the leaves of a tree were softly rustling. She looked up it and saw a large animal sprawled on one of the branches. It had the air of a man lying at ease in his hammock. However, this animal was bigger than any man she knew, and rippling with strength. His golden coat was mottled with dark patterns like flowers.

Jaguar.

His gaze was completely steady. When he spoke, his voice was deep as the earth, a warm, throaty rumble: 'I've been admiring you for a while. I wondered when you would see me.'

Strangely, she was not afraid.

'Wait,' he said. 'I'll come to you.'

He leaped down and stood before her for a long moment. Then he then began to pace around her in a languid circle, speaking again: 'Woman, I have seen your misery. Let me improve your life. Come with me. You have nothing to lose and everything to gain. I am an excellent hunter. I will bring you the best meat every day. Marry me.'

She did not know what to say.

He stopped pacing and waited.

'You have nothing to fear. Let me prove it: touch me. I promise I will not bite.'

She reached out, trembling at first. When she finally dared rest her fingers on his fur, it was soft and warm, muscles pulsing steadily underneath.

'Believe me, I will never treat you as prey. I will be gentle with you. I would like you to accept my offer, but I will not force you.'

'If I do …?' she whispered.

'I will teach you everything you need to know to live as a jaguar-woman. You will no longer have to keep company with the fellow humans who despise you. You will not be confined to a hut or limited to daytime. The whole forest and all the time, light and dark, will be yours. You will share my power.'

He fixed her for a long moment with his deep amber eyes, whiskers twitching. Then he shook himself lazily and strolled away into the trees. He did not look back.

By now the young woman was more afraid of missing her chance than of him. She called after him hoarsely, 'Wait, I am coming!'

The jaguar stopped just long enough for her to catch him up, then led her into the thickness of trees.

For several days, her brothers and their wives did not realize she was missing, assuming she had gone to stay with another family. They were all relieved, for they considered her to be a great nuisance. But when they asked around and were told that no one had seen her, they realized she had vanished. Being her closest kin, they were under strict obligation to find out what had happened to her. If she was in danger or had been abducted, they must rescue her. If she was dead, they must bring back her body and carry out funeral rites.

So, resentfully, they set out in search of her.

First, they followed the regular forest tracks that everyone used. There was no sign of her. Then they tried the narrower tracks that only the men passed down when they were hunting. They did not really expect to find her there, but on their way they met a troop of monkeys, running along the branches, chattering loudly.

'Where are you going to, men?' they called.

'We're seeking a young woman, our sister, who's gone missing,' the men replied.

'Another human? We've seen her, we watched her!' the monkeys said, pointing. 'Walking that way! Following a jaguar! You'll find her with him. Good luck – you'll need it!'

So the brothers went along the track that the monkeys indicated. The sun set behind the trees, and the sky above the canopy darkened. Stars shone alongside the moon.

As they walked, they kept calling out her name, on and on through the night.

And at last – they heard her calling back to them!

'Brothers, so you finally bothered to look for me, did you? No doubt you did it out of duty, not for love. Well, there's no need to worry because I've left you for good now. Don't shed any tears.'

'Where are you?' they called.

'I'm up in the trees.'

'Why are you hiding up there?'

'I'm not hiding,' she said. 'I'm resting.'

'Come home, you fool,' they scolded her. 'If you need a rest, use a hammock, not a tree. You'll be savaged by dangerous beasts up there.'

'I won't,' she said. 'You don't understand what you're talking about. My husband's taught me how to be comfortable here – much better than I ever was in *your* company.'

The brothers consulted each other in low tones.

'She's gone mad,' said one.

'From starvation and getting lost in the forest,' the other agreed.

'We'll have to ask the shaman to heal her after we bring her home.'

She called down to them again. 'My hearing's far better now than it used to be. I know everything you just said. I'm still perfectly sane. And I'm really content here. My new husband brings me more delicious meat than I can eat. Thanks to his lessons, I'm a hunter too now, like you – but far more skilled than you will ever be.'

In the moonlight, they saw a shadow moving down a nearby tree trunk. As it slunk through the undergrowth towards them, they saw what had become of their sister.

No longer a young woman but a jaguar, lithe and proud.

The two brothers stared at her, awe-struck and fearful. Then they fled.

WORLD WITHOUT END

The Wild Ones Fight Back

Portugal

Many people had abandoned their affinity with the other beings who shared their beautiful world. They derided their fellow creatures as mere 'things', devoid of either intelligence or feelings, and treated them with great callousness. They stigmatized some species as 'vermin' and tried to exterminate them.

Those who had been tamed and domesticated had no choice but to accept this state of affairs. But it was a different matter for the wild ones.

King Robin of the birds called a meeting to discuss how he and his fellow wild creatures could unite and take action. On the appointed day, hundreds of them – feathered, furred, scaled and invertebrate – congregated under a tall, gnarled tree. From his nest in the lower branches, King Robin warmly welcomed them. When everyone had settled down, he opened the meeting as follows:

'My friends, there are now far too many humans, and they have overrun the world. It would be impossible to hold them all to account. I therefore suggest that we choose an easy and obvious target as a showpiece, to make it plain that we no longer tolerate their domination.'

A babble of grunts, squeals, tweets, barks, snorts, caws, roars and so on signalled his fellow creatures' unanimous agreement.

King Robin went on, 'I have in mind a particular young male human, who many of you are already aware of. He is constantly setting painful traps in which to catch us. Not because he needs us for food (which would be excusable) – but simply because he enjoys tormenting and torturing us. Over recent years, countless victims have endured unspeakable suffering at his hands. So I suggest we humiliate and molest him in a similar way. We must carry out this punishment in a public place, to ensure that all humans hear about it. Then perhaps they will come to respect our feelings, and also defer to our intelligence.

What do you all think?'

He was answered at once by a loud buzzing, as a magnificent striped insect flew up to alight on the rim of King Robin's nest. 'May I speak?' she asked.

'Of course, Queen Bee,' King Robin replied respectfully.

'I know exactly which young human you mean,' she buzzed. 'He would be a most appropriate target. For every year he and his father invade my hive. They blatantly steal the honey that my poor workers have laboured all summer to make and cause dreadful damage. I would like to offer my tribe's unbridled support in carrying out this punishment.'

There were many exclamations of agreement, followed by a long discussion over exactly what form the wild ones' revenge should take. This produced numerous suggestions which, one by one, were all dismissed as unworkable.

Finally, Fox stepped forwards, holding up his paw. 'Friends,' he said, 'my tribe has also long suffered from humans constantly attacking us. We can't even settle down to enjoy a quiet meal without being chased out of our dens. They accuse us of overbreeding our young to become thieving pests, and have us beaten, torn apart or shot ...'

At this, a rat called out, 'Pests? Overbreeding? That's what they say about us too. Humans kill far more of us than of you. And as we all know, it's really humans themselves who cause all the world's problems.'

'True!' other voices agreed.

'Exactly!'

'Let's show those humans we're just as good as them.'

'*Better* than them!'

Fox blinked his shrewd yellow eyes around the gathering. 'What I suggest is this. We set a trap to capture this young human who will be our showpiece. We all know the cruel technique he uses on you unfortunate birds: traps made of gummy stuff – birdlime, I believe the humans call it. He tempts you by setting it next to your favourite food, and as soon as you land there, your feathers get stuck. Then you must struggle and thrash about, desperately trying to get free, tearing the plumage from your poor bodies, covering yourselves in wounds and bruises. Well, I propose that we set a similar sticky trap to catch *him*.'

'Yes, oh yes!' came a chorus of voices.

'But what form of trap would be effective to lure the young human?' someone asked.

Fox gave a scream of laughter. 'All those humans care about is enriching themselves with treasure,' he said. 'So let's shape the birdlime to look like a chest of gold coins. As soon as the young one sees it, he will try to either carry it away, or wrench the lid open. Either way, he will immediately find himself stuck fast to it. When he shouts for help, that will bring more humans – who will also get stuck!'

A great cheer went up, for everyone could immediately see this was a brilliant scheme.

Queen Bee said, 'My tribe can supply wax to mix into the birdlime. But we are far too small and fragile to build the trap itself: you larger ones must take on that job.'

'Will anyone volunteer?' asked King Robin. 'But don't put yourself forward unless you know how to avoid sticking to the birdlime yourselves as you work it.'

At once a chorus of monkeys spoke up:

'We have clever hands – we can do it!'

'We'll smear and smooth them with olive oil, so we don't get stuck.'

'We'll start on it at once.'

And so it was done. The monkey troop was a large one, and it took them no longer than a single night to chew up a large pile of dark, sticky fruits, thicken them with beeswax, then shape the mixture into a huge, solid cube. Next, they used their long fingernails to etch lines and patterns into it. Their decorations made it look exactly like a wooden treasure chest, complete with lid, lock and handle. When it was finished, they placed it just outside the house where the young human lived.

By that time, the sun was rising. All the wild creatures hid in the surrounding woods, waiting to see what would happen.

Very soon, the young human came outside. When he saw the birdlime 'treasure chest', he ran excitedly up to it and tried to open the lid with one hand – which immediately stuck to it. He tried to break free by leaning on it with the other hand – but this stuck too. He kicked it – and thus his foot became stuck also.

'Help!' he shouted.

At once other humans came running out of his house, followed by more from the surrounding houses. They attempted to pull the young one free, but to no avail. So they tried wrenching the birdlime box away from him – but only succeeded in getting themselves stuck to it too. Very soon a mass of people were writhing and struggling there.

The wild ones all came sneaking out of their hiding places and gathered around to watch. How they relished seeing their enemies suffering the very same cruelty that they themselves had invented! And how humiliating it was for the trapped humans to see the wild ones' mocking eyes on them!

With a loud, eager buzzing, Queen Bee offered to let her swarm finish off the victims by stinging them to death. However, the other creatures agreed with King Robin that it was preferable to let them struggle for as long as possible. This enabled even more humans to come and witness how the tables had been turned. They were extremely shocked to see the wild animals wielding such power.

At last the wretched humans in the sticky trap stopped struggling. They fell still and silent. A pack of wolves trotted up, dug a set of deep holes nearby and carefully buried the human remains there. Then all the birds, animals and insects drifted away to attend to their usual daily business.

The wild ones had made their point.

The Earth Fights Back

Mesopotamia (ancient Iraq)

'In those days,
the world teemed and the people multiplied.
The world bellowed like an angry bull.
So the gods agreed to exterminate humankind …

With the first light of dawn
a black cloud came from the horizon.
The storm god turned daylight to darkness,
and smashed the land like a cup,
sending a wail of despair up to heaven.
The tempest raged, gathering fury as it went,
pouring over the people like the tides of battle.
For six days and six nights the winds blew.
The warring armies of torrent, tempest and flood
raged together, overwhelming the world.

When the seventh day dawned,
the storm subsided.
The sea grew calm and the flood was stilled.
There was silence.
The surface of the sea stretched as flat as a roof-top.
On every side was the waste of water.'

The Earth Reborn

Ancient Norse people (Scandinavia)

At last –
the earth rose again from the ocean!
So green it was now, everywhere so green,
sparkling with waterfalls.
Eagles flew above,
hunting fish between the mountains.
Without even being sown,
crops began to ripen in the fields.

Golden playing pieces lay scattered in the grass:
poignant reminders of games
played and lost
by the long dead, mighty ones
of ancient times.

All past harms were permanently healed.
New people came to tame the wild lands,
building houses and halls
more splendid than the sun.
They lived in righteousness and peace,
blessed with joy for ever more.

Flowers Bloom Again

Yuwaalaraay Euahlayi people (Australia)

How desolate the Earth looked now! What a barren landscape! Sand, mud, rocks and water: there was nothing to uplift the heart.

What could be done?

The only ones who might find a solution were the wise ones, the clevermen.

A group of these sages sat quietly apart. They searched deep through their ancestral memories, mulling over treasured traditions. Images grew slowly in their heads, of fragrance and colours. In their minds' eyes they saw petals softly unfolding, quivering in the breeze and reaching for the sky.

They resolved to search for them.

Secretly, telling no one else, they set out together, walking fast to the northeast: an interminable journey. At last, their way was blocked by a precipitous mountain, its peak lost in the clouds.

They stared at it, praying. And their prayers were answered. Holes suddenly became visible in the rocks, shaped like steps, one above another, winding upwards.

The clevermen formed a line and began to ascend the steps. They climbed for a whole day, clinging precariously to the mountain face. They climbed for two days, they climbed for three. On the fourth day, they reached the top.

They found themselves on an undulating summit. The ground in front of them cracked open and a spring came gushing up. How thirstily they drank from it! Their weariness faded away.

All around them roared the wind. The roaring turned into a sonorous buzzing, the sound of a bull-roarer. Through the bull-roarer, they heard an invisible voice – a messenger from the spirits.

'What do you want here?' the voice called to them. 'What sacred knowledge are you seeking?'

One of the clevermen replied on behalf of his fellows, 'We have come to request flowers to make the Earth bloom again – as we know it did long ago, in the time of our ancestors. We want to bring joy back to the people, and to all other living creatures.'

'Wait,' said the voice.

They waited.

The buzzing spun itself into a whirlwind. It lifted the clevermen and carried them even further up, away from solid ground, on and on, then through an opening in the sky.

At last the whirlwind calmed and set them to rest in a dazzling land. It was completely covered in flowers.

They grew in lines, they grew in circles, they grew in coils and swirls. Some crept along the ground. Some spiked upward. Others were dotted about on scented bushes. Some colours were brilliant, others subtle. They were shaped like stars and flames, like clouds and rainbows.

The clevermen were overcome by their beauty. They stared and wept and laughed and hugged each other.

But the spirit voice chided them, 'Do not forget why you are here, do not delay. There will never be another chance like this ever again. Gather as many blooms as you can carry between you. When all your arms are full, you must stop. Then you will be conveyed back to the path that leads you home.'

So the clevermen reached up, they stooped and fell to their knees, urgently plucking the blooms. These they twisted behind their ears and into their hair. They draped them over their shoulders, clutched them between their lips, filled their hands and arms with them.

At the very moment when all the clevermen found they could hold no more, they felt the whirlwind stirring again. It spun them round, lifted them, moving them through light and emptiness –

Until they found themselves standing again on the summit of the mountain.

For the last time, the invisible voice spoke through the bull-roarer: 'You have fulfilled your purpose in coming here. Now heed this promise to the people, and to all the beings who live beside you: Never again will the Earth will be so bare as it was when you began your journey. Though these flowers will change and fade and die with the seasons, each year they and their seeds will throw up new shoots. The winds will bring showers to nurture them and the sun will bless them. More than this: the flowers will feed the bees, and from their nectar the bees will make honey to sweeten your lives.'

Then there was only silence.

The clevermen made their slow descent of the steep mountain steps. Then they walked the long paths back to their waiting people.

All through their journey, the flowers stayed as fresh as when they had first gathered them. As they went, the clevermen scattered handfuls of them here and there, letting the wind catch them up and carry them far into the distance. Some fell on tree tops, some on the plains, some on ridges, some on the hillsides.

And in all these places, the flowers have continued to grow ever since.

The Cycle of Renewal

Aztec people (ancient Mexico)

Many times in the past, Planet Earth has faced a cataclysm and survived it.

Aeons before written history, people lived under the gaze of the First Sun. Then monstrous beasts emerged and devoured them all.

After that, the First Sun burned itself out and vanished.

But it was not the end.

For a Second Sun rose in the sky. Under its light, new people were born. They multiplied and flourished, until a violent windstorm blew up. This storm swept everything away: all the buildings, all the trees, all the animals and all the people.

It swept away the Second Sun too.

But it was not the end.

For next a Third Sun was born. Once again, people thrived and came to dominate the world. But eventually, the sky caught fire. Its flames lashed down on the Earth below, destroying everything and everyone, leaving nothing but ash and cinders.

It was this fire that swallowed up the Third Sun.

But it was not the end.

For a Fourth Sun appeared, vitalizing a new community of people. Then suddenly the sky exploded with torrential rain. This caused such deep floods that everyone drowned.

There was such a deluge, so much water, that the Fourth Sun completely dissolved in it.

But even that was not the end.

For eventually, a new light appeared in the sky, welcoming life with its brightness and warmth: the Fifth Sun.

It was born of the sacred ones that came before, alive with movement, part of the cycle that never ends.

It is strong, beautiful and invincible, this one.
It is the very same Sun that shines on us today.

Accept the Ways of Nature

Ethiopia

All things live according to their nature.

The grass grows to live
and man burns it to live.

The river flows to live, and overflows its banks
because that is its nature; it cannot help it.
And man grieves when his planted fields are flooded,
for they are his life.

The tree cherishes its branches
because they are its beauty.

And the snake eats whatever it finds,
for that too is his nature.

So since all things act according to their nature,
let us dance and sing in thanks
because all things are as they are.

The Stream of Life

Tunisia

Far away in the mountains there was a spring of water. Flowing out of this spring came a sparkling stream.

This stream was restless and curious; it wanted to see and experience the whole wide world. So as soon as it escaped from the rocks where it was born, it started running. Fast and free it ran, determined that nothing would ever make it stop. It hurtled down the mountainside. It went on through deep valleys and shadowy gorges. It burbled and sang its way through thick forests. It sunned itself as it flowed across wide open plains. It went quietly through towns and cities, slinking beneath bridges, avoiding drains and houses. No obstacle could hinder it; not a boulder or a tree root or a boat or a boot or a road. So many countries it saw! So many different kinds of people and animals, all talking in a host of different tongues! Its knowledge of the world expanded every day. It felt very pleased with itself indeed. Surely nothing on Earth could rival its ever-growing wisdom!

Then one day the stream found it had reached the edge of the desert. In all its endless journey, it had never before come across such an inhospitable place.

There was nothing there except sand. And there was nothing to the sand at all, for it seemed to have no substance. It was soft as fog, yet dry as dead bones. It did not attempt to block the stream's way, or force it back, yet it refused to support its weight. It seemed to want to devour the stream, to totally destroy it. Over and over, the stream attempted to run onto the sand and across it. But each time, it felt its water being engulfed by an invisible force that made it vanish between the dry grains. If it carried on challenging the sand for too long, there would be nothing left of that stream at all.

So it drew to a halt on the desert edge, dribbling tentatively forwards, then hastily retreating like waves on a seashore. And as it lingered there, an invisible voice called out to it, saying, 'Greetings! I am the sand. Why are you disturbing me, stream? What do you want?'

'In order to fulfil my destiny, I want to cross your desert,' the stream called back. 'But that seems to be impossible.'

'Impossible?' the sand retorted. 'What nonsense! The wind regularly passes over my desert, so why shouldn't you?'

'Oh, it's all very well for the wind,' the stream said wretchedly. 'For the wind can fly. But I can't, for I am burdened with the weight of my water. Usually when I meet an obstacle, I dash myself against it, splashing and drenching it until it gives in and lets me pass. But I've already tried that many times with you – and failed.'

'Of course you have failed,' the sand chuckled. 'As you have seen, to repel you, I simply swallow you up. Don't let me take too much of your water – or I shall transform you to a swamp! Then you will never flow anywhere, ever again. But there is a way that you could cross my desert unharmed. Next time the wind comes flying this way, ask it to carry you over.'

'That's a ridiculous idea,' said the stream. 'How could the wind carry a heavy mass of water like me?'

'Oh, it's easily done,' the sand replied. 'All you have to do is relax and make yourself still, so the wind can absorb you.'

'Absorb me?' the stream exclaimed in horror. 'But if it did that, my individuality would be completely lost. I would no longer be the world-famous stream that has travelled further and seen more than any other living being. How would I find my true self again?'

'Ah,' said the sand. 'I see you overestimate your knowledge and understanding. I assure you that the wind has long been an expert in such matters. I regularly let it carry water across my desert. Each time, when it reaches the far side, it releases the water to fall as rain – which gathers on the ground and turns back into a river.'

For a long time, the stream was silent. Then it said, 'You have already caused much harm by obstructing me, sand. Why should I trust you to tell the truth?'

'Well,' said the sand, 'you know what choice you have. Either take my advice – or get stuck in me and degrade yourself into a mucky quagmire.' It lowered its voice. 'Why can't you see what is best for you? But wait! What perfect timing, for your chance has come!'

The next moment, the air began to stir. The stream felt something touching its surface and rippling its waters. It was not an unpleasant sensation. Indeed, there was something strangely familiar about it, an echo of long forgotten, ancient days.

'Ready yourself – for here it comes!' cried the sand.

Yes, it was the wind.

It hovered over the stream and whistled down to it: 'Do you not remember me, old friend?' Its tone was benign and reassuring. 'You may have hardly noticed me, but I have already carried your droplets through the air many times before, always taking great care of you. So come again into my embrace – come!'

The stream recognized the timeless wisdom of the wind's words, and yielded. It released its water as a cloud of vapour and let it float up. The wind absorbed it into its welcoming arms. Then, guarding it like a newborn infant, it carried the water right across the expansive desert to the far side. There the wind blew the water even further, to a distant mountaintop, and released it.

Gently, the water tumbled back down. Slowly, it gathered itself together – and reformed into its old shape, a sparkling, running stream.

'Welcome home, old friend,' the wind said softly. 'Look, you are completely free to resume your endless journey.'

In this way the wind taught the stream something it had never stopped to consider before, for it had been far too busy. This invaluable lesson is just as important to us today – we who yearn for proof that our world will never die:

The stream of life will always continue its journey,
for its way is written in the sands.

About the Stories in this Book

This book contains myths, legends, folk tales and fragments of wisdom from all over the world. My aim is to share global insights about how humanity can come to terms with the environmental crisis which affects us all.

Some of the stories were originally told by small-scale indigenous cultures, others come from large dominant societies, with many variations in-between. They are all treated equally for the purposes of this book.

Myths, legends and folk tales all have their roots in oral traditions. By their very nature, they have no named original authors. They have developed and evolved through countless oral and sometimes written retellings over the centuries, and in many cases there is no single 'correct' version. Some were first shared over a hundred years ago by communities which historically did not use writing. So they recited their tales to trusted outsiders, who then wrote them down for posterity. These sources range from brief summaries in note form only; to verbatim translations from different languages, often reworked by the original collector to be more easily understood. The transcribers usually came from beyond the narrators' own cultures, working within a colonial context which had a very different ethos from today. Nevertheless, their work in recording the stories as accurately as possible is invaluable. Others were first written down many centuries ago in ancient manuscripts; or were documented relatively recently in more modern story collections and academic papers.

Most of the sources I used can now be viewed online, through websites that specialize in making out-of-copyright books and journals available to the public, and also in print editions. I have made every effort to find the earliest versions, and to respectfully convey their special flavours and meanings. It is the intention of this book for their world-class wisdom to continue inspiring people everywhere to take better care of the natural world. For information about how I have retold each story, please see the specific notes in the pages that follow. These also name the relevant sources, and give interesting background information.

Beginning (page 10)

Martin Boord, 'Tibet and Mongolia' in Roy Willis (Gen. Ed.), *World Mythology* (London: Duncan Baird Publishers, 1993)

This myth belonged to an ancient shamanist faith practised in Tibet before Buddhism replaced it in the 8th century CE. Boord, a renowned scholar of Tibetan religion, does not give the source for his brief description of it. He says that the wandering shamans of ancient Tibet used to enter a trance to make contact with their numerous gods, demons and spirits. These lived in three realms: heaven, the Earth and the underground labyrinths.

Shamanism is a type of religion once widespread not only in Asia, but also in parts of Africa, the Americas and the Pacific islands. It involves specialist practitioners interacting with the perceived world of spirits for clairvoyance, healing, practical help and other purposes.

Creation (page 12)

Alice Marriott and Carol K Rachlin, *American Indian Mythology* (New York and Scarborough, Ontario: New American Library, 1968)

The Cheyenne people were originally farmers in the indigenous cultural area of North America known as the 'great lakes and woodlands'. In the 18th century they were forced onto the Great Plains, where they adopted the local nomadic culture. The men hunted buffalo from horseback, while the women processed the meat, hides and body parts into food, clothing, artefacts and tipis, and gathered edible wild roots and berries. Today, they are split between two indigenous nations within the USA: the Cheyenne and Arapaho Tribes of Oklahoma, and the Northern Cheyenne Tribe of Montana.

The source book authors collected this story directly from a Cheyenne woman called Mary Little Bear Inkanish in 1960. Some elements of her account were slightly muddled, but I have followed it closely in terms of content, style and imagery.

She named the male All Spirit as Maheo. However, throughout this book I have deliberately omitted specific names, in order to convey a feeling of unity between stories from very diverse cultures.

A comment near the end of the original narrative says that the myth explains why:

> 'Grandmother Turtle and all her descendants [i.e. the turtle species] must walk very slowly, for they carry the weight of the whole world and all its peoples on their backs.'

Some modern Native American peoples today still refer to either the world, or North America, as 'Turtle Island', and there are parallel myths from neighbouring peoples.

Growth (page 17)

Antony Alpers, *Legends of the South Sea* (London: John Murray, 1970)

Tahiti is a small, mainly mountainous island in the central Pacific Ocean, fringed by coral reefs and lagoons. It is part of an island group now known as Polynesia, which also includes New Zealand, Hawaii and Easter Island. Long before formally recorded history, its people were already skilled navigators and sailors, making long journeys across vast expanses of ocean. They lived by growing food plants, and hunting birds and fish, and produced splendid handcraft items. Their culture was based on both social and religious hierarchies, with elaborate rules of etiquette. Their rich oral literature on both religious and secular themes, was expressed through poetry, prose, song and chants.

In the late 18th century, the island was visited for the first time by European whaling ships and merchants from the penal colonies of Australia, followed by Christian missionaries. These newcomers brought alcohol, weapons, new diseases and turmoil, gradually overwhelming the time-honoured way of life. Today, Tahiti is part of the semi-autonomous territory of French Polynesia, with its own assembly, president and laws. Around 70 per cent of the current population is of Polynesian ancestry, and still celebrate their ancient culture through the performing arts.

The poem here contains extracts from their complex and dramatic creation myth, narrated mostly in prose with short sections in verse. It tells how the creator, Tangaroa made the land, growing things, water and living creatures, then fixed the sky dome on its pillars. The complete myth was transcribed while the indigenous faith still held sway, in two versions: the first narrated by four traditional priests in 1822, the latter in 1835.

Seasons (page 18)

Donald Alexander Mackenzie, *Wonder Tales from Scottish Myth & Legend* (London: Blackie and Son, 1917)

'Wee White Hoose', *Older than Time: The Myth of the Cailleach, the Great Mother* (weewhitehoose.co.uk/study/the-cailleach, 2021)

The narrative traditions of Scotland are usually associated with lighthearted folk tales rather than myths; but since the early 20th century, folklorists have unearthed more serious stories from oral traditions that seem to have pre-Christian origins. This story is one of them, pieced together by collectors from seasonal customs, local place names and archaeology.

In some sources and quasi-academic commentaries, the winter witch-goddess is called 'the Cailleach', which is said to translate as 'old woman' or 'hag'. Mackenzie names her as Beira, Queen of Winter. The Wee White Hoose website ('exploring stories, traditions, and folklore from Scotland') says she is known throughout the British Isles by various names, some in the local Celtic languages' including 'the Blue Hag of Winter' and 'Bone Mother'.

Mackenzie names her son who appears in this story as Angus, saying that he has a number of brothers who are all gigantic and quarrelsome. He says Angus's lady is called Bride.

The changing seasons, and particularly the fear of severe winters, is also a common theme in Native American and Arctic mythology.

Good and Evil (page 24)

www.mapuche-nation.org

The Mapuche are a large group of indigenous people whose territory extends across parts of southern Chile and neighbouring Argentina. Estimates of their current population vary from 1.3–2 million people. Many of their members continue to value and, in some cases, practise a lifestyle, beliefs and traditions very different from those of mainstream society.

Their ancestral religion is closely linked to the land and the natural environment; in fact, their very name means 'Earth people'. The fragment of wisdom presented here is taken from a description of their spiritual beliefs on an official website of the Mapuche Nation. These also teach that no living creature could live without the grace of the great spirit, and that human beings are an integral part of nature.

Their concept of good and evil is further explained in a paper on the website of Universidad de Chile. This relates the two opposing forces to the cardinal points. Good is represented by south and east, since the sun, moon and stars rise there, generating life. The north – bringer of bad weather – and the west – where the sun sets and the dead are said to rest – are both linked to evil.

The constant battle between good and evil is a major facet of religion and mythology throughout the world.

The collection of stories that follow in this section portray some of the environmental evils that the world is currently confronting.

Man-made Climate Disaster (page 25)

Cyril Birch, *Chinese Myths and Fantasies* (New York: H Z Walck, 1961)

The mythical dragons of China – immortal beings who mingle with the deities – appear in numerous traditional tales. They are very different from the evil, human-eating, fire-breathing European creatures that share their name. Although Chinese dragons can be volatile, vindictive and dangerous, they also play a vital role in maintaining the Earth's environmental balance, and sometimes support human characters in distress. Most importantly, their element is not fire but water. Thus, they are strongly associated with rainfall, and also with rivers and streams, lakes and the sea. Many stories tell of them causing catastrophic floods.

These watery domains are said to be ruled by powerful dragon kings. They are huge and terrifying in appearance, with yellow scales, beards hanging from long snouts, hairy tails, gaping sharp-teethed mouths and blazing eyes. They typically live in coloured stone and crystal palaces deep under the water, surrounded by priceless treasures, sometimes alongside several generations of their families. Some stories describe them as attended by servants and military personnel in the form of fish and other sea creatures.

Chinese dragons can be of either sex, as demonstrated in this story by the dragon king-mother. Female dragons tend to be more sympathetic to ordinary people than the males. In fact, a number of legends tell of youthful dragon princesses who come to the rescue of young men, or even sacrifice their own lives to save a human community from trouble inflicted by a fierce dragon king.

In Chinese folk religion, the Emperor of Heaven (or Jade Emperor) rules over Heaven, Earth, the Underworld and all the important deities. Heaven is traditionally portrayed as a replica of the earthly bureaucracy, with departments and civil servants responsible for particular domains and functions.

In the source book, Birch names the hero of this story as a warrior called Li Ching, who later became the imperial Duke of Wei. His version of the story concludes with the dragon king-mother saying the heavenly authorities have punished her by 80 strokes of the rod, for enabling the disastrous flood. She expects her absent sons to be disciplined in the same way for neglecting their duty. However, she generously forgives Li Ching, accepting that a mortal could not properly understand the art of rainmaking. In fact, she rewards him with a bag full of pearls. He uses these to buy food for the inhabitants of the village he had inadvertently flooded, until they can reclaim their land and rebuild the village.

Global Warming (page 31)

James Mooney, *Myths of the Cherokee* (Washington, D.C.: Bureau of American Ethnology, 1902)

The Cherokee people, descended from village-dwelling farmers, are today based in Oklahoma and North Carolina. They are the largest Native American group in the USA.

The story related here is part of a longer, quite complex myth. After the Sun burns the inhabitants of Earth with her hot rays, the people ask supernatural beings known as the Little Men for help. They transform into snakes, intending to kill the Sun with a poisonous bite; however, by mistake they kill the Sun's daughter instead. The distraught Sun goes into hiding, not just cooling the world but also turning it dangerously dark. The Little Men now advise the people to fetch the Sun's daughter in a box from the Land of Ghosts, warning them not to open the box until they reach home. Unfortunately, they give in to the daughter's entreaties to let her out on the way. She transforms into a bird, increasing her mother's grief even more. Eventually, some young people perform a dance which persuades the Sun to smile down on Earth again.

This ending of a failed attempt to bring someone back from the land of the dead is a common mythical motif worldwide. You can read the full story in my book, *Native American Myths*.

The Sun is a common character in the myths of other Native American peoples, but is usually a powerful masculine deity, while the Moon is portrayed as female.

Wildfire (page 34)

William D Westervelt, *Myths and Legends of Hawaii* (Honolulu: Mutual Publishing, 1987)

Politically, Hawaii is a state within the USA. However, geographically it lies in the Pacific Ocean 2,000 miles (3,200km) from the mainland. It comprises a chain of 137 islands extending for 1,500 miles (2,400km), some of them still actively volcanic. The culture of its indigenous population is Polynesian.

Westervelt was an American immigrant to Hawaii, who became one of the greatest experts in the islands' traditional culture and oral literature, researching and publishing this work – much of it collected directly from local people – between 1900 and 1925.

He names the god of fire as Ai-Laau, which translates literally as 'the one who devours trees or a forest' and says his home was in the crater of Kīlauea, in the south of Hawaii Island (also known as the Big Island). Kīlauea is still an active volcano, with regular – though often minor – eruptions. There do not seem to be many proper stories recorded about Ai-Laau, perhaps because he relinquished his home and status among the volcanoes to the goddess Pele. She is a major and very colourful figure in the archipelago's mythology.

Climate change and global warming are contributing to an increasing number of devastating wildfires around the world. At the time of writing (2023–24), they were causing particular problems in Canada, Greece, Amazonia, Chile – and, indeed, in Hawaii itself.

Drought (page 35)

Georgi Kapchits, *Drought in Somali Folklore* (Finnish Somalia Network: Horn of Africa Journal, 2012. See online article at afrikansarvi.fi/issue3/38-kolumni/97-drought-in-somali-folklore)

During the Middle Ages, Somalia was a major commercial centre dominated by powerful emperors, and producing significant poets, writers and Muslim scholars. Following centuries of European colonization and internal turmoil, it is currently one of the least developed countries in the world. Political difficulties are exacerbated by its hot, arid climate, with regular failed rainy seasons and water shortages. In 2023, it was estimated that a recent drought had brought famine to 3 million Somali people, with tens of thousands dying, and hundreds of thousands displaced from their homes. Meteorologists predict that man-made climate change will cause more droughts in the future, afflicting many parts of the world.

Kapchits is an expert on Somali language and folklore. He gives very brief outlines of several traditional narratives on this subject, which he found in a local collection called *Xikmad Soomaali* (Somali Wisdom) published in 1956. My retelling is based mainly on one of the animal fables he describes, blending in the soothsayer episode from another drought story.

Traditional tales with animals as the sole protagonists are very popular throughout Africa. The jackal – a species of wild dog – often features as a trickster. In real life, the jackal is notorious for stealth, cunning and adaptability, and for its supposed ability to conceal its tracks and even feign death.

Famine (page 40)

Sophia DeLeon, *The Féar Gortach: Memorialization of the Great Hunger in Irish Folklore* (University of Florida Journal of Undergraduate Research, 2023. See online article at https://journals.flvc.org/UFJUR/article/view/133160)

Leanbh Pearson, *Irish Folklore: Féar Gortach* (see online article at: https://leanbhpearson.com/2023/05/01/irish-folklore-fear-gortach)

Famine is usually associated with the developing world, but until relatively recently it could also afflict western Europe – particularly, in Ireland in the mid-19th century.

At the time of writing, the Republic of Ireland (which excludes the six counties of Northern Ireland) is among the wealthiest countries in the world. However, until 1922, the entire island of Ireland was a largely impoverished part of the United Kingdom, with most of its people living as tenant farmers on smallholdings owned by disinterested, absentee English landlords. The single staple food of the Irish population was potatoes, but in 1845, almost the entire crop was infected by an airborne disease known as blight. This continued until 1852, causing widespread famine, death from starvation and emigration. It is estimated that about 1 million people died and millions more fled – more than a quarter of the total population. This period is known as the Great Famine or the Great Hunger.

On a stark political level, there were many avoidable reasons for this widespread suffering. However, just as in the modern world, some people looked for more sinister, populist causes. Ireland has a rich tradition of folklore about supernatural beings, and it was from these roots that stories of the Hungry Man developed.

The Great Famine is an important part of Irish history, and the Hungry Man must have featured strongly in informal oral storytelling during the 19th and early 20th centuries. However, I could not find any transcriptions of actual stories about him. So my retelling here is based on colourful folk beliefs recorded in several research articles, and descriptions on Irish cultural websites.

Linked to these were tales of the Hungry Grass. These tell of healthy young people becoming lost in broad daylight on particular patches of supposedly 'cursed' grassland. Some die on the spot from unnatural hunger; others are found in a confused state and gradually waste away from starvation. It was said that the Hungry Grass was haunted by ghosts of earlier victims, waiting to drag others onto it to suffer the same fate.

Litter (page 44)

Charles J Finger, *Tales from Silver Lands* (New York: Doubleday & Company, 1924)

Finger says that, while travelling through South America, he was told this story by an elderly indigenous man who talked at great length deep into the night, 'picking out a piece of the tale here and there, and explaining until I was well nigh like to get the story tangled myself'. In modern Colombia, about 10 per cent of the population identify as indigenous, split between 102 different groups.

The story is actually the first part of a much longer narrative. In the second section, the lazy villagers are confronted by a stranger who clears up all their mess, then offers to work for them without reward, enabling them to continue their life of idleness. It sounds too good to be true – and indeed, it is. The stranger gives each village family a set of 20 wooden mannikins to carry out all their daily chores. However, the villagers are soon overwhelmed by their ceaseless activity, so they beg the mannikins to stop work. In response, the mannikins attack the villagers, forcing them to flee. Finally, the people realize they must take responsibility and work for themselves. Meanwhile, the mannikins are transformed to the first monkeys – which take revenge on the people by bombarding them with nuts.

This latter section seems to be based on ancient Moche (Peruvian) and Mayan (Central American) myths which describe the revolt of man-made objects and domestic animals against humans. These have been interpreted from curious images on ancient pottery vessels and murals, depicting objects such as headdresses, weapons and shields with arms and legs, chasing people or holding them captive. There are also written records of indigenous Peruvian and Mayan myths which describe domestic utensils and animals rebelling and turning the tables against their callous human masters.

Polluting the Sea (page 48)

Olive Beaupre Miller (Ed.), *Tales Told in Holland* (Chicago: Book House for Children, 1926)

This is based on a well-known Dutch legend about a real place, 'The Lady of Stavoren'. According to Miller, Stavoren was once a mighty and very wealthy city with a splendid harbour full of thriving merchant ships. She concludes her story by saying that the sandbar – known as 'Lady's Sand' – can still be seen, and says that the once great city has become a sleepy village. Even today, a hundred years later, Stavoren still has a relatively small population of under a thousand.

Harming Animals (page 52)

Richard F Burton, *The Book of a Thousand Nights and a Night, A Plain and Literal Translation of the Arabian Nights Entertainments*, Volume III (London(?): Privately printed by the Burton Club, 1885–88)

This story comes from the huge collection of ancient Arabic, Indian and Persian folk legends popularly known as *The Arabian Nights*, believed to have been compiled between the 8th and 13th centuries. They are held together by a dramatic 'frame story'. This describes a king finding his wife committing adultery with a servant (or slave). He takes revenge against all womankind by forcibly marrying one new bride after another, then executing each one the following morning. However, his last victim, Shehrazad, changes his behaviour by telling the king a series of exciting stories every night, pausing each one on a cliffhanger just as dawn is breaking. The king is so desperate to hear the conclusions that he spares her life. You can read Shehrazad's own story in my book *Dark Fairy Tales of Fearless Women*.

All the stories in *The Arabian Nights* are narrated in Shehrazad's voice. Some have many layers of stories-within-stories – which is the case for this particular tale.

It opens with a peacock and his wife giving sanctuary to a terrified duck. The duck tells how she once dreamed of a voice warning her to beware of human wiles and traps. Greatly troubled, she wanders around until she meets a young lion who has similarly been warned to avoid humans. She follows the lion to the crossroads, where she observes the adventure described in my retelling. In the version narrated by the duck, the carpenter pushes the box containing the lion into a pit and sets light to it. After she finishes her story, the peacock and peahen assure the duck she is safe with them. However, their idyll is destroyed when a ship anchors at their island home, and the crew come ashore. The peacock and peahen manage to escape; but the duck is paralyzed with fear and gets carried off to the ship.

Like all *Arabian Nights* stories, the original text emphasizes the Muslim culture from which they originated. In this tale, humans are called 'Sons of Adam', there is discussion of predestined fate, and it ends with the peahen declaring that the duck died as punishment for not sufficiently praising Allah.

War (page 59)

Richard Erdoes and Alfonso Ortiz, *American Indian Myths and Legends* (New York: Pantheon Books, 1984)

Erdoes says this story was narrated to him in 1969 by a 70-year-old Sioux medicine man called John (Fire) Lame Deer in Rosebud Indian Reservation, South Dakota. The Sioux people belong to the Great Plains culture of nomadic buffalo hunters (see the Notes to Creation, page 159). They comprise seven closely related tribes known as the Seven Council Fires (*Oceti Sakowin*), today based in both the USA and Canada. When this story was collected, the people of Rosebud still practised many of their traditional sacred ceremonies, and the modern Rosebud Sioux Tribe website says that the whole premise of their society continues to be based on respect. Their land comprises over 900,000 acres (364,000ha) of grassland scattered with pine forest, deep valleys, steep hills, ravines and lakes. It is home to 20 communities.

The thunderbird is one of the iconic supernatural characters of Native American and Canadian First Nation stories. In 1909 it was described by Charles Eastman, a renowned Sioux doctor, writer and reformer, as follows:

'The Great Bird of storm and tempest, who was appointed in the beginning of things to keep the earth and also the upper air pure and clean. Although there is sometimes death and destruction in his path, yet he is a servant of the Great Mystery and his work is good.'

You can read more about thunderbirds in my book *Native American Myths*.

Extinction (page 62)

Jan Knappart, *African Mythology* (London: Diamond Books, 1995)

The story was one of a huge collection obtained by Knappart when he conducted fieldwork in different parts of Africa between 1957 and 1988. He says it comes from Gongola, a former administrative region of northern Nigeria, which is now split between the states of Adamawa and Taraba. The region is home to a number of different peoples, but he does not name the ethnic group to which this story belongs.

In pre-industrial societies around the world, hunting was not a sport but an essential source of meat. Knappart says that in some traditional sub-Saharan African communities, successful hunters were highly honoured as major suppliers of food. It was commonly believed that magic powers could be used to bewitch game animals, which thus willingly allowed themselves to be killed.

The scene at the end of this story – in which the hero and the animals shape-shift into a variety of forms, as part of a chase – is a common motif in myths and legends worldwide.

The Earth is Our Mother (page 68)

James A Teit, 'Okanagon Tales', in Franz Boas (Ed.), *Folktales of Salishan and Sahaptin Tribes* (Lancaster, PA, and New York: American Folk-Lore Society, 1917)

These words form the introduction to a creation myth narrated in the early 20th century by an elder of the Nespelem people called Red-Arm (KwElkwElta'xEn). Similar sentiments are echoed in other Native American myths which describe the Earth as the mother of both nature and human beings.

The traditional territory of the Nespelem was along the banks of a river in Washington that shares their name. They are now one of the Confederated Tribes of Colville Indian Reservation in the same state.

This lies within the cultural region known as 'the plateau', spanning the northwestern USA and parts of British Columbia in Canada. Historically the plateau people lived in small groups, migrating with the seasons over large areas to gather wild roots and berries, to hunt and to fish.

Beware of Consequences (page 69)

Maurice Russell, *Told to Burmese Children* (London: The Epworth Press, 1956)

This traditional tale was collected by a British teacher of English who worked in Myanmar (at that time known as Burma) during the 1950s. He indicates that he originally learned the stories he presented from the young girls and boys who were his school pupils.

According to legend, Burma was ruled by kings from as far back as the 9th century BCE. It remained a kingdom until its conquest by Britain in the 19th century, and it is now a republic.

Value Everything (page 71)

Alex Ruelas, *6 Maya Myths and Legends that will Show You a World of Ancient Magic* (Magazine Maya Luxe, 2021. See online article at: https://www.magazinemayaluxe.com/articles/maya-myths-and-legends-ancient-magic)

The Maya are an indigenous people of Central America, currently numbering some 8 million. They are descendants of an ancient civilization of the same name (*c.*250–1697 CE), and still live in their ancestral lands of modern Mexico, Guatemala, Belize, El Salvador and Honduras. Some maintain their old cultural traditions, while others have adopted modern urban lifestyles.

This story appears in a Mexican tourism online magazine. The author does not give its source, and it is not clear whether it is an ancient narrative or a modern fable – possibly both. However, he says, 'These stories are alive, still present in daily folklore'.

The ciricote tree (*Cordia dodecandra*) does indeed have multiple uses. In Mexico the fruit is commercially processed into a dessert known as *dulce de ciricote*, made into jams and preserves, and used to feed pigs. The rough leaves and their sap are still valued for cleaning cooking pots, and carpenters also use the leaves as a substitute for sandpaper. Traditional medicine uses infusions brewed from the bark for treating diarrhoea, dysentery, asthma, bronchitis and coughs. The wood itself, being hard and resistant to humidity, is very durable and is used for many kinds of building work and handcraft.

Recent research shows that up to 90 per cent of trees in the Maya lands have potential use to humans, but unfortunately, Mexico has one of the highest rates of deforestation in the world.

Live in Harmony with the Wild (page 74)

Annu Jalais, *Bonbibi: Bridging Worlds* (Chennai: Indian Folklife, National Folklore Support Centre, 2008)

The location of this tale is the Sundarbans, the world's largest mangrove forest. It straddles river deltas and coastal regions of southern Bangladesh and north-east India above the Bay of Bengal, intersected by tidal streams and channels. Alongside the forests and barren mudflats some of the land is used for agriculture, and local villagers also work at fishing, crabbing, gathering wild honey and collecting and processing forest resources. Several sections of the Sundarbans are UNESCO World Heritage Sites, and it is an important habitat for Bengal tigers and hundreds of other species. However, the entire eco-system is under threat.

There are a number of outline accounts of this legend and its associated beliefs on Indian academic, tourist and popular websites; and many shrines in the area display statuettes of the Tiger-Demon and Bonbibi. However, I could not find any proper written narratives, suggesting that it mainly exists in oral folklore tradition. Jalais says that its written sources are a late 19th-century Bengali text called *Johuranama*, and a 17th-century epic poem, *Ray-mangal*.

Inspired by this and associated legends, local people have great respect for nature. This includes tigers, despite their occasional deadly attacks. Workers endeavour to enter the forest only during permitted times of day, and avoid it at night out of respect for nocturnal wildlife. They do not carry firearms, and refrain from smoking and defecating there. Fish must not be caught during the breeding season, and neither wood nor honey must be taken from small flowering trees. A local proverb declares:

Facing any danger inside the forest,
whoever prays to her,
Mother Bonbibi protects them all.

The Sundarbans are home to both Hindu and Muslim communities. Each believes in Bonbibi, adorning the stories and beliefs with their own religious details, and encouraging a spirit of mutual tolerance.

Clear Up Your Own Mess (page 81)

Harold Courlander (Ed.), *Ride with the Sun: An Anthology of Folk-Tales and Stories from the United Nations* (London: Edmund Ward publishers Ltd, 1957)

Courlander says that he discovered this moral tale as an oral rendering in Israel by a man called Joseph Meltzer, who based it on a section of the *Talmud*. This is a book of ancient laws and wisdom which has long served as the primary guide for Jewish life and thought.

One of the scandals of the modern world is a phenomenon sometimes called 'waste colonialism', in which western countries export huge amounts of plastic and other waste to developing countries. The intention is for recipient countries to recycle this, and thus earn valuable foreign income by participating in international trade. However, much of the waste ends up in landfill, is burned, or simply litters the environment.

It appears there were similar antisocial dumping problems in the ancient world, albeit on a much smaller scale. For the *Talmud* was originally compiled as long ago as the 5th century CE, in Babylonia (modern Iraq). This section of its text opens with the statement: 'The sages taught: a person should not throw stones from his property into the public domain.'

Leave Some for the Future (page 84)

Ida Ayu Laksmita Sari and I Nyoman Darma Putra, *Narrative on Nature Conservation: A Comparative Study of the Folktales of Bali Aga and Ainu* (Malaysia: Kemanusiaan, The Asian Journal of Humanities, 2020)

The Ainu are an indigenous people who live on the island of Hokkaido in northern Japan. Their ancient history and customs set them apart from mainstream society, but in the 18th century they were forced to assimilate. After long campaigning, their separate identity was formally recognized by the Japanese government in 2008, though without legal protections. Recently there has been increased interest in their culture, though partly as a tourist attraction.

An important theme of Ainu myths and legends is the interdependent relationship between human beings and nature. This is linked to their belief that everything in the world is inhabited by a spirit: not just living beings and plants, but also water, wind, fire, mountains and so on. This means that they strive to treat everything with respect, holding that the spirits are ubiquitous and constantly vigilant.

Don't Waste Food (page 86)

Ella Elizabeth Clark, *Indian Legends of Canada* (Toronto: McClelland and Stewart Limited, 1960)

Clark says she discovered this story in a book called *Wandering Round Lake Superior*, published in 1860.

The Ojibwe (Chippewa) people's homelands extend across the indigenous cultural area known as the northeastern woodlands, straddling parts of southern Canada and the Northeastern United States. They are currently the second largest Canadian First Nations group, and one of the largest Native American peoples in the USA.

Historically they were semi-nomadic, living in wigwam villages comprising a number of families, and moving with the seasons to carry out a wide variety of activities. These included gathering wild rice and other foods, fishing, hunting and harvesting wild maple syrup.

It was the women who planted small fields of corn (the cereal known in the UK as maize), beans and squash – often called 'the three sisters'. Numerous different varieties of corn grow in North America, ranging in colour from pale yellow to dark red and blue. It was ground in a pestle and mortar for bread, or used whole in stews and soups. Corn was an important crop, and there are many other interesting Native American myths about it, often describing its supernatural origins.

Take Responsibility (page 90)

Ida Ayu Laksmita Sari and I Nyoman Darma Putra, *Narrative on Nature Conservation: A Comparative Study of the Folktales of Bali Aga and Ainu* (Malaysia: Kemanusiaan, The Asian Journal of Humanities, 2020)

This story is described in the same academic paper as 'Leave Some for the Future' (see Notes on page 172).

The Bali Aga people are an ethnic minority group who live in the mountainous eastern and northern regions of Bali, one of the islands of Indonesia. They still maintain many aspects of their traditional way of life. In recent years, their villages have been heavily promoted by the government as tourist attractions.

With its reference to commercial logging, this story clearly has fairly modern origins. It was collected in 2017, from a narrator called I Wayan Sukrata in the village of Pedawa in northern Bali. Widespread deforestation for commercial purposes has long been a major environmental problem throughout Indonesia. This is particularly concerning since the country has some of the most biologically diverse forests in the world, and is home to a huge number of species.

Unlike most of Indonesia, the principal religion of Bali is Hinduism. Monkeys there are considered to be sacred, as is the forest they inhabit. The main species is the long-tailed macaque, which can often be spotted mischievously interacting with people around the local temples.

Protect the Water (page 93)

www.caribbeanreads.com; www.tntisland.com/folklore

Mama D'Lo appears in a number of accounts of Caribbean folklore from St Lucia, but particularly from Trinidad and Tobago. However, she receives less prominence than the terrifying characters that dominate most of the region's traditional stories. Lurking in pools high up in the mountains, she is the guardian of all water creatures. She protects them from people who pollute the water, and from those who kill without good reason.

In some tales she is married to Papa Bois – once a formidable hunter, but now protector of the forest plants and animals. He is elderly, ragged and very hairy, half-man/half-goat, with horns and cloven feet. He carries a bamboo horn, which he sounds to warn animals that humans are near. He can run faster than any other being, and shape-shift into a large animal in order to frighten people away. It is said that anyone who encounters him should be very polite and avoid staring at his hooves.

Though Mama D'Lo's name is derived from the French for Mother of Water, her origins seem to lie in widespread African beliefs in the water spirit Mami Wata. She too is often described as being seen combing her hair in the form of a half-woman/half-snake; or as fully human but with a snake wrapped around her. Although Mami Wata has a number of important supernatural roles, she was not considered a nature guardian until enslaved people from West Africa carried her across the Atlantic to the Caribbean. Here they developed her into the supernatural being known in the islands today.

Keep the Song Alive (page 94)

Colin M Turnbull, *The Forest People* (London: Chatto & Windus, 1961)

The Mbuti (Bambuti) people live in the Ituri Rainforest in the north-east of the Democratic Republic of the Congo. Their current population is estimated at 30,000–40,000 people. Both women and men are hunters and gatherers, trading some of their food with outsiders for agricultural produce and manufactured goods. During the rains they live in small villages; but when the dry season starts, they move to camps deeper inside the forest. Their society is informal and largely egalitarian, with both sexes discussing important matters and reaching decisions together. Their traditional way of life is at risk due to incursions into the forest from mining, deforestation and plantations. It is also threatened by persecution, conservation policies which prohibit hunting of large game, and civil unrest in the wider country.

The Mbuti are intrinsically linked to the forest which is both their home and their source of food. It is said that they perceive it as a living, breathing entity – not to be worshipped in the manner of a god, but to be respected and protected, anticipating punishment for misuse. Newborn babies are clothed in forest products and bathed in forest water, while older children are taught not to insult the forest by unnecessary noise, destruction or uncivilized behaviour.

Songs are equally important in many aspects of Mbuti life. They are usually sung by a group of people, sometimes along with dancing and other rituals. Song is used to awaken, celebrate and make peace with the forest to ensure its continuing support. It is also used for instruction, to bring luck in hunting and gathering, and in times of trouble and grief.

In the Time of the Ancient Ones (page 100)

pib.socioambiental.org/en/Povo:Aparai

The Aparai are an indigenous people of the Amazon Basin rainforest. Their homes are mostly in Brazil, but also in Surinam and French Guiana. Numbering only a few hundred, they live in scattered small villages, alongside their closely related neighbours the Wayana people. Historically they practised hunting, gathering and 'slash-and-burn' agriculture. Since the 20th century they have also produced handcraft items to sell to outsiders.

The fragment of wisdom given here is based on their accounts of a mythical time when, they say, there was little difference between human beings and animals. According to their stories, sometimes the different species all cooperated and helped each other, while at others there was conflict and competition among them. It was from the animals that people learned important knowledge such as hunting techniques, healing, songs, dances and art.

Similar beliefs and stories were also common among many Native American and Canadian First Nation peoples.

Animals Have Feelings Like Ours (page 101)

J D Lewis-Williams (Ed.), *Stories that Float from Afar – Ancestral Folklore of the San of Southern Africa* (Cape Town: David Philip, 2000)

The San people live mainly in Botswana, with smaller populations in Namibia, Angola, Zambia, Zimbabwe, Lesotho and South Africa. Historically, they formed small family groups of hunters and gatherers, moving seasonally to follow the availability of water, edible plants and prey animals. Periodically, these groups would gather together to socialize, exchange news and arrange marriages. They had hereditary chiefs, but most important decisions were made by group discussion. Since the 1950s, many San people have become farmers, encouraged by the Botswana government.

This story was originally collected in 1876. It was narrated by a man called Dia!kwain, who said he had first heard it from his mother, ≠Kammi-an. (The symbols within the names represent the sounds of the San people's 'click' language.)

Baboons are among the world's biggest monkeys, and very common in Africa. They live for up to 40 years in groups of between 20 and 300 animals, spending most of their time on the ground, eating an omnivorous diet. The young are entirely dependent on their mothers until they are about three or four months old, when they start to find their own food. Baboons are currently suffering habitat loss due to overgrazing, agricultural expansion, irrigation projects and human settlement. Regarded as pests, they are widely killed. They are also hunted for their skins and held captive for experimentation in laboratories.

Never Hurt a Fellow Creature (page 104)

Alanson Skinner and John V. Satterlee, *Folklore of the Menomini Indians* (New York: Anthropological Papers of the American Museum of Natural History, Vol. XIII, part II, 1915)

The Menominee (Mamaceqtaw) people belong to North America's 'northeast woodlands' cultural area and traditionally lived by hunting, gathering, fishing and small-scale crop farming. Their historical lands covered some 10 million acres (4 million ha) in what is now Wisconsin and Upper Michigan in the USA. Their website states:

> 'We have managed to keep a fraction of our ancestral territory for a home which is now our reservation. We continue to have

> strong leadership ... that has taken us through much adversity. We continue to speak our language and practice our traditions ... Spiritually, we continue to speak with our creator through tobacco, prayers, and other offerings ... We are a sovereign nation ... that refused to be pushed from our territory, a nation that will remain strong and independent.'

In his introduction to the source book, Skinner writes that he collected the stories during five summers between 1910 and 1914, speaking to many Menominee people who still remembered the old way of life. He would sit informally in their homes, listening to the stories 'told just before bedtime'. He says that some casual tales 'have for the most part come up in ordinary conversation', but in order to hear the more formal myths and legends, 'I have had to pay, as a Menomini youth must do.' The practice of payment indicates the serious status accorded to such stories. He names several informants, both women and men, most importantly his 'adopted uncle, Judge Sabatis Perrote, in whose cabin I often stayed.'

This story reflects the people's ancient belief in the inner spiritual powers of people, animals and natural phenomena such as storms.

Porcupines are solitary and mostly nocturnal, often resting in trees. The coat of an adult contains about 30,000 quills – modified hairs that form sharp, barbed spines, for both insulation and defence. They are normally kept flat against the body but can easily be pulled out. When threatened, the quills stand erect, and a mere swing of the tail is enough to dislodge them towards an attacker, releasing hundreds in a single encounter. Another defence is a strong, unpleasant smell, which increases when the porcupine is agitated. Dyed porcupine quills have long been used by many Native American peoples for creating elaborately decorated textiles, such as clothing, boxes, moccasins, cradleboards, etc. The craft, unique to North America, is known as quillwork.

Dogs Show How to Cooperate (page 108)

Emilie Demant Hatt, *By the Fire – Sami Folktales and Legends*, translated by Barbara Sjoholm from the original Danish edition, *Ved ildn Eventyr og historier fra Lapland*, published in Copenhagen, 1922 (Minneapolis: University of Minnesota Press, 2019)

The Sami are an indigenous people of northern Europe. Their homeland, called Sapmi (historically known to outsiders as Lapland), is spread across the mainly Arctic regions of Norway, Sweden, Finland and the Russian Kola Peninsula. There are currently estimated to be about 80,000 Sami people, of whom at least 20,000 live in Sweden. Although stereotypically associated with reindeer herding, only about 10 per cent of modern Sami people work in this field, which has been greatly modernized with the use of snowmobiles, all-wheel-drive vehicles and helicopters. Many have now integrated into mainstream urban culture. Others work at small-scale farming or fishing, sometimes supplemented by handcraft items, which are in particular demand for the tourist market. Following many centuries of prejudice and suppression from wider society and national governments, official attempts are now being made to improve their status and welfare.

Hatt collected the stories in the source book in Arctic Sweden between 1907 and 1916, during which time she periodically lived with two families of Sami nomadic reindeer herders. She shared their everyday life, slept in their summer tents and travelled with them on foot. The story retold here was one of two about dogs that she heard from a woman called Margreta Bengtsson. Hatt gives a vivid description of how the reindeer herders used to enjoy storytelling:

> 'When the darkness draws them to the campfire, when the stew kettle hangs on its sooty chain and steam and smoke rises through the tent opening to the clouds and night sky, then rest comes, memories slip in like dreams to a sleeper ... The faces of young and old light up with smiles of recognition when the right strings are plucked and the music comes; one tells a story, while another adds an important detail. The words spill out, sometimes fantastic and lurid, yet the finest poetry and the purest magic come through as well.'

Arctic foxes are found in northern Europe, northern Asia and North America. They have been extensively hunted for their thick, pure white fur, and also suffer from the effects of global warming. At the start of the millennium, they were almost extinct in Sweden, Norway and Finland. However, since 2020, aided by official conservation programmes, their numbers have greatly increased. Wolverines are the biggest members of the weasel family, living all over the Arctic regions. They feed mainly on carrion but are also powerful hunters capable of bringing down reindeer. Sweden is currently the main stronghold for wolverines in Europe but lists the species as vulnerable.

A common breed of Sami reindeer herding dogs is the Finnish Lapphund, which is ideally hardy and energetic, but also calm and friendly. It is medium-sized, with pointed ears, and usually has black or dark grey fur with lighter markings.

People and Elephants Helping Each Other (page 111)

E B Cowell (Ed.), *The Jataka or Stories of the Buddha's Former Births*, Vol. II, translated from the Pali by W H D Rouse (Cambridge: Cambridge University Press, 1895)

The *Jataka* is a collection of hundreds of legends and fables relating to previous incarnations of the Buddha, founder of the Buddhist religion. They depict him as taking either human or animal form to illustrate virtues such as generosity, thus setting an example for followers to try and emulate. The stories generally open with an introduction explaining the Buddha's reason for telling it; and end with spiritual explanations of the characters.

There is some speculation about their antiquity, but depictions in ancient Indian art indicate they were probably being told well over 2,000 years ago. The Cowell translation that I used says that the Pali language text dates to at least 380 BCE, and 'includes many stories which have travelled afar. Many can be traced cross-culturally in the folklore of many countries.'

This story comes from a section known as *Alinacitta-Jataka*. As in my retelling of this story, the original text features some of the elephants' own thought processes; indeed, it presents much of the plot mainly from the elephants' perspective. It concludes with the white elephant growing so fond of the king that adopted him, that the courtiers delay telling him of the king's death until the queen is able to introduce him to her own young son – who is the latest incarnation of the Buddha. The elephant immediately shows high regard for the new prince, and then goes on to frighten away enemy troops that had been attempting to capture the kingdom. Nothing is told about what happens to the old bull elephant after he has handed over to his son.

Captive elephants, with their extraordinary strength and exceptional intelligence, are believed to have been used as beasts of burden in India and Southeast Asia for over 4,000 years. Although the elephant in this tale voluntarily offers his services to the carpenters, today it is considered very cruel to train and use such animals for work. Through the ages – and to a much lesser extent today – they have been forced to carry both goods and people, and even to transport soldiers and weapons in battle. They are also exploited for human 'entertainment', for example giving rides to tourists and performing degrading tricks. In India, some are kept on short chains in temples where they are believed to bring good luck and forced to take part in processions and ceremonies. Some countries such as Thailand and Myanmar have now banned using elephants in commercial logging; but this raises further problems with caring for the numerous elephants already held in captivity, who can no longer earn their keep.

Learning to Live with Mouse Neighbours (page 117)

Charles F Lummis, *Pueblo Indian Folk-Stories* (New York: Century Co, 1910)

The Native Americans known as 'the Pueblo People' live in the Southwestern United States – a land of huge, dramatic landscapes with mountains, flat-topped mesas and weather-carved rock formations that rise above deserts and sweeping stretches of scrubland. The pueblos are ancient towns that were originally built hundreds of years before the first white settlers arrived, and are still occupied today. The oldest sections, often built on a mesa top, comprise rows of permanent, multi-storey housing blocks constructed of adobe or stone, with different levels connected by ladders. Each block contains a number of separate apartments, typically built around a central courtyard. An important space within the pueblo is the *kiva*, a large, circular underground room, used for traditional religious ceremonies and worship.

This story comes from Isleta Pueblo in New Mexico, just south of Albuquerque. It was established about 700 years ago and is one of the biggest pueblos in the state. Historically, the people were farmers, mainly growing corn (maize); today, the pueblo owns over 200,000 acres (80,000ha) of land and its most profitable businesses are cattle ranching and various entertainment complexes. It is currently home to over 3,000 people.

In his introduction to the stories, collected around the turn of the 20th century, Lummis says that winter was the season for storytelling, which was a popular activity enjoyed by all ages. He heard this particular tale from a 'withered' elderly man called Diego, before an audience of eager young boys. He concludes his narrative by saying, 'That is why we have never been able to drive the mice out of our homes to this day.'

Wild mice are found in virtually every country and every kind of habitat in the world, both natural and man-made. They voraciously eat anything they can find, and in domestic settings also chew wires, paper, insulation and other items for nests. Females start breeding at between four and seven weeks old, and one can produce over 50 young every year. As this legend demonstrates, it is impossible to totally eliminate them.

Loyalty to a Whale (page 120)

Sir George Grey, *Polynesian Mythology*, originally published in the Māori language in 1854 (Christchurch: Whitcombe and Tombs Limited, 1965)

The Māori people inhabited New Zealand before European colonization. However, they themselves did not settle there until around the 14th century, apparently arriving in canoes from other Polynesian islands. They lived in small communities, maintaining a strong oral literature about warfare, their mythical ancestors, and gods of the natural world. Europeans arrived in the early 19th century, bringing huge changes with the introduction of literacy, overseas trade and the Christian religion. The country became a British colony in 1840, gaining full independence in 1947. By the mid-20th century, many Māori people had become urbanized. After campaigns to restore their rights and culture, their language is now taught in some schools, there are Māori TV and radio channels, and a number of Māori people are members of the national parliament. Their population is currently estimated to be around 800,000. The official Māori name for New Zealand is Aotearoa, often translated as 'Land of the Long White Cloud'.

The source book seems to be the earliest written record of their traditional narratives. Grey was colonial Governor-in-Chief of New Zealand in the 1840s, and one of his major tasks was to deal with the Māori people's grievances against their British occupiers. He realized he could only do this effectively by studying their culture, particularly as the Māori chiefs often quoted fragments of ancient oral narratives during negotiations. So, Grey spent eight years learning the language, and being instructed in myths, legends and proverbs by influential chiefs and priests.

Almost half the world's whale and dolphin species are found in New Zealand's waters. In pre-European times, Māori people did not hunt them, but harvested meat from whales that occasionally stranded on their seashores, alongside bones and ivory to make weapons and ornaments. It was not until the 20th century that industrial-scale whaling began there. However, since the 1970s such activity has been opposed by many local people, and in 1991 a proposed global ban on commercial whaling became official government policy.

Wolves as Protectors (page 124)

Thersa Matsuura, *The Uncanny Japan Podcast* (uncannyjapan.com/podcast/okami-japanese-wolf-part-2/)

John Knight, *On the Extinction of the Japanese Wolf* (Nagoya: Asian Folklore Studies, Vol. 56, No.1, 1997)

The story comes from a summary on the website of Matsuura's *Uncanny Japan* podcast. She says it is based on a folk tale in a collection published by Koyama Masao in the 1930s. Matsuura mentions other tales of wolves protecting people who treat them with kindness and respect, and this theme is explored in more detail in Knight's academic study of the Kii Peninsula in southern Honshu, Japan's largest island.

Wolves can adapt to many different kinds of environment and once ranged over most parts of the northern hemisphere. Currently the largest populations are in Russia and Canada. They can also still be found in many parts of Asia, Europe and the USA, with varying status and numbers. Their survival depends not just on having sufficient food and habitat, but also on human acceptance.

In Japan – as in many other countries – historically they were demonized and decried as rabid killers. Systematic hunting and poisoning, encouraged by a bounty system, led to them being declared extinct there in 1905. However, Knight wrote some 90 years later, that many people across the country still insisted they had encountered wolves 'in their youth'. Some of these informants did not claim to have actually seen a wolf; instead, they had experienced a strong sense that one was present, perhaps – improbably – hiding behind a single reed nearby. Others had interpreted the chirping of a sparrow to be the form into which a wolf had magically transformed. Knight summarizes a number of other legends that demonstrate wolves' benevolence, including some that tell of infants abandoned in the wilderness, then found and raised by wolves. They are also said to howl in particular places to warn people of imminent natural disasters.

Today there are grassroots efforts to reintroduce the animals through the Japan Wolf Association.

Learning from Bears (page 128)

C M Barbeau, *Huron and Wyandot Mythology,* with an appendix containing earlier published records (Ottawa: Government Printing Bureau, 1915)

The Wyandot (Wyandotte, Wendat, Huron) people have a complex history. They were originally an alliance of nations around the north shore of Lake Ontario, farming and trading in corn and tobacco. In the 17th century they were defeated and dispersed by the Iroquois confederacy, with some moving southwards into what is now the USA. Today the multi-ethnic Wyandotte Nation of nearly 7,000 members is based in Oklahoma. Their website states: 'We adopted many white captives into the Nation. Many obtained high tribal status and made significant contributions to the betterment of the tribe.' In Canada, the mostly urbanized Huron-Wendat Nation is based in Quebec, with about 5,000 members.

In the introduction to his collection of stories, Barbeau says that they were collected directly from dictation by mostly named people. Sometimes this was in the indigenous tongue, sometimes in English or French if the narrators could speak one of those languages. Even in cases that needed to be translated, the collectors were careful 'not to alter the character of the material and, as far as possible, to retain the informant's style and expressions.' This particular tale was recorded by his fellow ethnographer, W E Connelley in 1899. He names a number of informants but does not specify who told him which story.

A traditional way of opening a Wyandot narrative was by using one of several idiomatic phrases stating whether or not it was believed to be 'true'. No such indication was given for this one. However, the origin of medicinal knowledge was obviously very important, and was probably part of the great canon of myths passed down through the generations. Such stories were generally recited during long winter nights by the fireside, for if told during the summer, it was said that toads or snakes would crawl into one's bed.

Bears are among the most common animal characters in Native American and Canadian First Nations myths and legends, often marrying humans, or adopting human children. The outcomes vary between disturbing, tragic or happy, depending on the region and the purpose of the tale.

The idea that animals could have knowledge of medicinal plants without any human intervention has recently been proved true. In 2024 scientists filmed a wild orangutan in Indonesia with a large open wound on his face. He chose a specific plant, chewed up its leaves and stem, applied the resulting liquid onto his cheek, then covered the wound with a 'poultice' of the chewed leaves. It stayed free of infection, and within a month it was fully healed. The plant he used was later identified as one already known to have anti-inflammatory and anti-bacterial properties; in fact, it is used by local people to treat malaria and diabetes. Similar behaviour was recorded in the same year among wild gorillas in Gabon.

Sharing a Jaguar's Power (page 135)

pib.socioambiental.org/en/Povo:Manchineri

The Manchineri people live in the Amazon rainforest in north-west Brazil, and parts of neighbouring Peru and Bolivia. Traditionally, they were hunters and slash-and-burn farmers; however, since the 19th century their way of life has been transformed by invasions of rubber extractors, miners, loggers and farmers. In 2022, a Manchinerian representative was among indigenous Amazonian leaders who campaigned at a United Nations gathering in New York against the ongoing destruction of their forest by some of the world's biggest companies and banks.

Jaguars are the third biggest cat in the world, found only in Central and South America, with over half of their population in Brazil. They are carnivores, eating a wide variety of animals. Their name derives from the indigenous Tupi and Guarani word *yaguareté*, meaning 'true, fierce beast who kills in one leap'. They are significant characters in many other stories from the region, usually with supernatural qualities. They transform in and out of human shape and sometimes marry humans, but also attack, kill and eat them.

The website that gives this Jaguar story quotes the son of a renowned shaman saying,

> 'My father told me ... that someone would go hunting, an uncle, a nephew, and he would send a jaguar to accompany him. So if he became lost, the jaguar would appear and say, "No the path is right over there!" The jaguar worked for the [shaman].'

This intriguingly echoes the Japanese story 'Wolves as Protectors' (see Notes on page 182) – showing how totally different cultures with no geographical connection, can develop very similar beliefs.

The Wild Ones Fight Back (page 140)

Charles Sellers, *Tales from the Lands of Nuts and Grapes – Spanish and Portuguese Folklore* (London: Field & Tuer, the Leadenhall Press, 1888)

In the original story, the boy who enjoys trapping and killing birds is the motherless son of a notorious robber who also plunders church treasures. I have adapted the beginning very slightly to suit the wider purpose of this book.

The amusing episode in which a series of characters get stuck to an object one by one is very common in traditional tales in many parts of the world. Known as the 'tar baby' motif, hundreds of different versions have been recorded, from Africa, the Americas, the Caribbean and India. Probably the best-known one is from the African American *Uncle Remus* tales about Br'er Rabbit and his friends, first published by Joel Chandler Harris in 1881. Similar tales from Europe tell of people sticking to each other as they try to pluck a golden feather from a goose, or remove an enormous turnip from the ground.

The idea of animals deliberately fighting back against human exploitation and destruction may seem unlikely. However, while I was writing this book, several reputable news sources reported that whales had been attacking boats – apparently deliberately – in the Strait of Gibraltar between Europe and Africa. A yachtsman shared a video on social media, showing his vessel targeted by a pod of five orcas. They circled the boat, then took turns to repeatedly and systematically ram it in different places. By the time the crew and passengers were rescued, the boat was half submerged with its stern in the air, and later sank. Hundreds of similar incidents involving orcas had been reported in the same area during the previous four years, and also off the Atlantic coast of Portugal and Spain.

The Earth Fights Back (page 144)

N K Sandars (translator), *The Epic of Gilgamesh* (London: Penguin Books, 1960)

This comes from the oldest surviving written story in the world, a narrative poem known as *The Epic of Gilgamesh*. It is believed to have been composed nearly 4,000 years ago, perhaps based on even older oral versions. It was unknown until the mid-19th century, when excavations at Nineveh, a city of ancient Assyria in Mesopotamia (modern Iraq) revealed about 15,000 fragmented clay tablets. They were inscribed in a variety of languages, using an ancient script known as cuneiform. From these, archaeologists and linguists painstakingly pieced the story together.

It tells how Gilgamesh and his beloved friend Enkidu slay a monstrous guardian of the forest. Gilgamesh then spurns the goddess of war. She retaliates by sending a deadly supernatural bull, which the two heroes destroy. In revenge, the goddess kills Enkidu. Gilgamesh, devastated at his friend's death, sets out to find the secret of immortality. Eventually he meets the sage Utnapishtim, who sends him on an unsuccessful quest to obtain the plant of eternal youth. The lines here are quotes taken from Utnapishtim's description of a cosmic flood, which he had once survived by following divine advice to build a ship large enough to accommodate his extended family plus select craftsmen, animals and grains.

It is very similar in essence to the story of Noah in the Book of Genesis from the Old Testament of the Bible, also written in the Middle East. However, flood myths are not limited to that area; indeed, they are one of the most widespread mythical subjects, particularly in the Americas. Sometimes a flooded world is part of a creation story – for example, see Creation, page 12. More commonly, the deluge is a punishment from the gods for bad human behaviour. Such stories usually feature a handful of people who survive by boarding a special vessel, or climbing a high mountain or tree. Sometimes they also save other creatures and useful items.

The Earth Reborn (page 146)

Mimisbrunnr.Info, Developments in Ancient Germanic Studies, *The Comparative Voluspa*, www.mimisbrunnr.info/comparative-voluspa (features various translations of *Voluspa* from Old Norse to English, 1836, 1866, 1883,1908, 1923, 1928)

Andy Orchard (translator), *The Elder Edda: A Book of Viking Lore* (London: Penguin Books, 2011)

This is my own rewording of various translations from the final verses of a mystical poem called *Voluspa* ('The Prophesy of the Seeress'), believed to have roots in the Viking Age (8th–11th centuries) or even earlier. Some scholars believe it may be several poems cobbled together. In summary it describes the creation of the world and its inhabitants, both mortal and supernatural; followed by conflict between the deities resulting in its destruction. It concludes with the world's rebirth, as presented here. The style of *Voluspa* suggests it was based on a series of dreams supposedly experienced and recounted by a wise woman with the gift of clairvoyance. It switches between the first person and the third person, with the tense likewise moving between past and present.

It is intriguing how neatly it links in to the previous mythical extract, which comes from a totally different civilization and time. It also points to the two stories that follow it, both from other, very distinct cultures.

The Viking Age was the period when merchants, plunderers, invaders and migrants originally from Scandinavia carried their culture westwards, eastwards and southwards. They established flourishing colonies, particularly in Britain, Ireland, Iceland and Greenland. *Voluspa* was written in the Old Norse language, which is closely related to modern Icelandic. It comes from a 13th-century book variously known as the *Codex Regius* ('Royal Manuscript'), the *Elder Edda* or the *Poetic Edda*. Though some pages are missing, the 29 poems contained in its 90 pages are an invaluable source of Norse mythology, transcribed from oral traditions which were circulating much further back in time.

Flowers Bloom Again (page 148)

K Langloh Parker, *Australian Legendary Tales*, based on the original collection published in London, 1896, selected and edited by H Drake-Brockman (Sydney: Angus and Robertson, 1953)

The Yuwaalaraay Euahlayi people's traditional lands covered parts of what are now southern Queensland and northern New South Wales. They were formally recognized as title holders over a section of this in 2021.

The original text states that the events of the story took place after Baiame (the 'Great One' or 'Creator') had left the Earth to dwell in the upper world. The voice that speaks to the clevermen on top of the mountain belongs to Walla-guroon-bu-an, Baiame's spirit messenger. The clevermen scatter the flowers they have brought back at a place called Girraween, 'The Place of Flowers'. Today, a national park in Queensland bears the same name, with its Aboriginal First Nations heritage clearly acknowledged in official information. There is also a suburb of Sydney, New South Wales called Girraween.

The Aboriginal peoples of Australia have roots believed to go back tens of thousands of years. Historically they formed over 400 separate communities, each with its own language or dialect, connected by trade and inter-cultural networks. Their religious and other traditional beliefs vary according to community, but are all are strongly connected to the land. They have been preserved through stories, songs and art, with traditional healers being important custodians. A major concept of their sacred stories is known today as 'The Dreamtime', a term invented by European anthropologists. It refers to the period of creation by spirit beings and ancestors, when human beings were taught to treat the Earth and other creatures with respect. It provides a spiritual structure and a set of rules for every aspect of life: social, economic, religious and ritualistic.

European colonization of Australia began when British ships brought convicted prisoners there in 1788, followed by voluntary immigrants from the early 19th century onwards. It became a British colony, then a fully independent country in 1942. During this period, the Aboriginal people were subjected to profound cultural and social oppression. Eventually, in the mid 1960s, they were finally granted citizenship and given the right of self-determination for their future. Slow progress is still being made towards their conciliation with the Australian white majority.

This story was one of many collected by Catherine Field in the late 19th century, published under her married name of Mrs Langloh Parker. She grew up among Aboriginal people living on her father's large livestock farm, and in adulthood continued to live as their neighbour in New South Wales. She believed her four books of Aboriginal myths and legends to be to be the first proper written collection of this material.

The term 'cleverman', now widely used in Australia, is the literal translation of the indigenous word *wirinun*. It refers to a highly respected healer and guardian of ancestral tradition, sometimes said to have supernatural powers. Clevermen may be male or female, depending on the community. Their knowledge is passed on by instruction, followed by special ceremonies.

The Cycle of Renewal (page 151)

John Bierhorst, *History and Mythology of the Aztecs: the Codex Chimalpopoca*, originally written in 1558, translated from the Nahuatl (Tuscon: University of Arizona Press, 1992)

The Aztec Empire was an alliance of three city-states that flourished in what is now central Mexico, from 1428 until 1521, when it was conquered by the Spanish. It was a richly developed civilization, based on a hierarchy of nobles, commoners and slaves. There was apparently relative sexual equality, with women owning property, and some working as merchants, doctors, priests, and midwives. For men, the activity which conferred the highest status was warfare. Aztec religion was centred on four gods: Tlaloc who ruled rain and storms; Huitzilopochtli who ruled the sun and warfare; Quetzalcoatl who ruled the wind, sky and stars; and Tezcatlipoca who ruled the night, magic, prophesy and fate. Numbers and calendars were important in religious belief and practice, following both a 365-day solar calendar, and a 260-day ritual calendar, which coincided only once every 52 years. Each day had a name and number in both calendars.

This is reflected in the original words of the myth retold here. It was written down in the 16th century in the local Nahuatl language, in a book known as the *Codex Chimalpopoca*. Fortunately, a photographic facsimile of this was made in 1945, since the original went missing four years later and has not yet been traced. The Aztecs and their predecessors had long been recording their history and sacred traditions in books of pictographs. The oldest ones were destroyed by their own leaders in 1430 for containing 'falsehoods'; while later ones were dealt with in the same way by the Spanish for being 'un-Christian'. It is thus particularly lucky that the *Codex Chimalpopoca* survived.

Bierhorst translates the opening lines, and the date of writing, as follows:

> 'Here are wisdom tales made long ago, of how the Earth was established, how everything was established, how whatever is known started, how all the suns there were began.
> There are 2513 years today, on the 22nd day of May, 1558.'

This particular myth is sparse, vague and mysterious, full of obscure references and numbers. It provides little information about anything except for the demise of each set of human beings and the sun that shone on them. Bierhorst suggests this is because:

> 'the author speaks to us as though we were looking over his shoulder while he points to painted figures ... In places the text reads like a sequence of captions, as though the unseen pictures could capture the burden of the tale.'

My retelling presents it in a simplified form, omitting the arcane numbers, names and other details.

The concept of a series of suns overseeing the birth, death and rebirth of the world was also important further south in the cosmology of the Inca Empire. This was centred in what is now Peru, extending into parts of modern Argentina, Bolivia, Chile, Colombia and Ecuador, before being conquered by the Spanish in 1532.

Accept the Ways of Nature (page 153)

Harold Courlander and Wolf Leslau, *The Fire on the Mountain and other Ethiopian Stories* (New York: Henry Holt and Company, 1950)

When Courlander and Leslau collected their stories in the 1940s, the country was mainly rural, with farms in the mountain valleys, and nomadic cattle and goat herders moving over the wide, grassy plains. There were just a few cities, such as the capital, Addis Ababa.

The Ethiopian population is made up of over 80 different ethnic groups. Despite this diversity of peoples and languages, the source book says:

> 'The stories the people tell around their fires at night are the same ... For they have been carried back and forth for centuries in the migrations of cattle herders, by camel caravans, and by the traders who cross the mountains with bracelets, knives and spears to sell.'

Indeed, the authors discovered that, though some traditional Ethiopian tales are purely local in both content and flavour, many are related to similar ones told in other regions of Africa, as well as in the Middle East and Europe. They typically feature both human and animal heroes and villains, wise and foolish, often exposing human weaknesses and vanities, or making moralistic points.

The fragment of wisdom here comes from a story called 'The Judgement of the Wind'. Courlander and Leslau collected two versions of it, one from a priest in the city of Gondar, the other from a Somali boy (30 per cent of Somalis live in Ethiopia) who they met in New York. They say that similar tales are told by other East African peoples; and there are also Indian, Indonesian and Filipino versions.

The full narrative tells of a fearsome snake fleeing some hunters. The snake persuades a farmer to hide it – but as soon as the hunters have gone, it seizes the farmer to kill and eat him. When the farmer protests, the snake retorts, 'Although you saved me, I am hungry – so I have no choice.' Together they ask first a tree, then a river, then some grass to pass judgement on whether or not it is ethical for the snake to eat the helpful man. Each in turn points out that although they provide vital resources for human beings, the latter always respond negatively, setting a bad example for other beings to copy. Finally, they meet the wind, who gives its slightly more generous answer, as quoted. The story ends happily, with the wind contradicting itself to please all sides. It tells the snake: 'Since it is your nature to eat man, eat man'; yet advises the farmer, 'As it's your nature not to be eaten, do not be eaten'. In the ensuing confusion, the farmer is able to flee safely back to his village.

The Stream of Life (page 154)

Idries Shah, *Tales of the Dervishes* (St Albans: Granada Publishing Limited, 1973)

The source book attributes this beautiful story to a man called Awad Afifi the Tunisian, who died in 1870, adding that it is well known in the oral traditions of many languages, particularly in Dervish communities.

Dervishes are members of a Muslim religious order known as Sufism, taking vows of poverty and austerity. Sufism is a mystical branch of Islam, which seeks to provide a spiritual path to wisdom and union with God. Dervishes practise various rituals on their spiritual quest, including formulaic recitation of sacred phrases, dancing and 'whirling'. Over the centuries they have played an important role in some Muslim countries, but today many aspects of Sufism are considered unacceptable within mainstream Islam.

Shah says that Sufi students would immerse themselves in stories such as this one, with their teachers helping them to understand the deeper layers of meaning. Some of the stories have also passed into general folklore.

About the author

'Rosalind Kerven, connoisseur of myths and folktales'
The Independent

Rosalind Kerven is an expert in world myths, legends and folk tales, which she has been collecting and retelling for many years. With an academic background in social anthropology, she has thousands of stories on file, from numerous countries and cultures.

Rosalind has written over 70 books, highly acclaimed by reviewers and published in 22 countries. She has worked for many publishers including B T Batsford, the British Museum Press, the British Library, Cambridge University Press, Dorling Kindersley, Oxford University Press and the National Trust, as well as her own small imprint, Talking Stone. She is an active member of the Folklore Society and has contributed to many of their events. She lives in the Northumberland National Park, where she spends her free time walking in wild places, observing wildlife and gardening organically.

Other books by Rosalind Kerven

Celtic Fairy Tales and Legends

Dark Fairy Tales of Fearless Women

Medieval Legends of Love & Lust

Native American Myths

Viking Myths & Sagas

Faeries, Elves and Goblins: The Old Stories

Arthurian Legends

English Fairy Tales and Legends